Wireless Channel Measurement and Modeling in Mobile Communication Scenario

This book delves into the fundamental characteristics, measurement techniques, modeling methods, and theories of wireless channels in mobile scenarios.

Unlike wired communication systems, which are more predictable, wireless communication systems are significantly affected by radio propagation and wireless channels. By investigating the mechanisms of wireless channels and measurement techniques, this book aims to better understand wireless communication systems in order to optimize the quality and design of wireless communications. The title covers key topics in the field, including basic theory of radio wave propagation and non-stationary channels, theory and method of time-varying channel measurement, measurement case analysis, wireless channel modeling theory and parameter extraction method, rail traffic channel measurement and modeling, and dynamic modeling and simulation method of time-varying channels.

This book is suitable for researchers and students interested in radio wave propagation, wireless channels, and mobile communication systems. It can also serve as a useful guide for technical professionals who have a background in mobile communication technology.

Ruisi He is Professor at the State Key Laboratory of Advanced Rail Autonomous Operation and School of Electronic and Information Engineering, Beijing Jiaotong University, China. His research interests include wireless propagation channels, railway and vehicular communications, and 5G and 6G.

Bo Ai is Professor at the State Key Laboratory of Advanced Rail Autonomous Operation and School of Electronic and Information Engineering, Beijing Jiaotong University, China. His research interests include broadband mobile communication, dedicated mobile communication, and artificial intelligence.

Wireless Channel Measurement and Modeling in Mobile Communication Scenario
Theory and Application

Ruisi He and Bo Ai

CRC Press
Taylor & Francis Group
Boca Raton London New York

CRC Press is an imprint of the
Taylor & Francis Group, an **informa** business

人民邮电出版社
POSTS & TELECOM PRESS

This work is supported by the State Key Laboratory of Advanced Rail Autonomous Operation and School of Electronic and Information Engineering, Beijing Jiaotong University, China.

First edition published 2024
by CRC Press
2385 NW Executive Center Drive, Suite 320, Boca Raton FL 33431

and by CRC Press
4 Park Square, Milton Park, Abingdon, Oxon, OX14 4RN

CRC Press is an imprint of Taylor & Francis Group, LLC

English Version by permission of Posts and Telecom Press Co., Ltd.

ISBN: 978-1-032-66912-0 (hbk)
ISBN: 978-1-032-67179-6 (pbk)
ISBN: 978-1-032-66979-3 (ebk)

DOI: 10.1201/9781032669793

Typeset in Minion
by codeMantra

Contents

CHAPTER 1

Introduction

1.1 BASIC PRINCIPLES OF RADIO PROPAGATION

The characteristics of wireless channels are usually the foremost aspect in any wireless communication system, and are also related to a series of problems in wireless communication engineering design. Wireless channels in mobile scenarios are complex and cannot be analyzed simply by using methods and conclusions of fixed wireless communications. The specific characteristics of mobile environment and scenario must be analyzed and investigated with the help of basic theory of radio propagation. Radio propagation, as a highly theoretical discipline, has gained fruitful results after more than 100 years' development. However, although some basic theories of radio propagation have been developed, there are many problems yet to be investigated in combination with specific applications and environments, especially propagation mechanisms and wireless channel characteristics in mobile scenarios. At the same time, radio wave propagation is a highly practical technology, and its research has been centered on application. Especially today, its development is driven more by demand of applications. Mobile communication is one of the most active and rapidly developing fields of communications, and is one of the science and technology fields that will have a significant impact on human life and social development in the 21st century. Investigation of radio wave propagation theory and wireless channels in mobile scenarios is gradually becoming a hot topic.

First, we introduce the basic theory and mechanisms affecting radio wave propagation. After James Clerk Maxwell's pioneering research, people have gained a basic understanding of radio wave propagation. The physical existence of electromagnetic waves was confirmed by Heinrich Rudolf Hertz in 1887. However, Hertz did not discover the practical properties of electromagnetic waves and believed that radio waves could not be used to carry speech because of the low frequency of sound waves and the poor propagation properties of electromagnetic waves. In 1894, Oliver Lodge used these theories to build the first wireless communication system, although its transmission distance was only 150 m. In 1897, Guglielmo Marconi succeeded in sending wireless signals from the Isle of Wight in England to a tugboat 18 miles (29 km) away. In 1906, Reginald Fessenden made the first transmission of voice and music using amplitude modulation. He modulated low-frequency signals

DOI: 10.1201/9781032669793-1

onto high-frequency electromagnetic waves for transmission, thus breaking through what Hertz called the low-frequency propagation limit. This method is still commonly used in today's various wireless communication systems. Today, the term "mobile wave propagation" includes all practical applications of wireless communications, regardless of whether the sender and receiver are mobile. It includes satellite mobile communications, maritime mobile communications, as well as cordless phones, traditional private mobile systems, and cellular systems. Among these, the development of "land mobile radio propagation" is of particular interest because of its use in everyday life [1].

The division of the large-scale and small-scale categories of radio wave propagation is a great advancement in the perception of the propagation characteristics of radio waves. The division is based on the size of the window through which people observe radio waves. The large-scale characteristics of radio wave propagation are defined based on its distance scale, reflecting the signal field strength variations over long distances (typically hundreds to thousands of meters) between the receiver and the transmitter. The study of large-scale characteristics focuses on the prediction of path loss and the estimation of shadowing. Broadly speaking, the large-scale characteristics include both path loss and shadowing. The former is caused by the radiated spread of the transmit power and the propagation characteristics of the channel; the latter is caused by obstacles randomly present between the transmitter and receiver through absorption, reflection, diffraction, and scattering. Small-scale characteristics describe the characteristics of rapid fluctuations in the received field strength over short distances (a few wavelengths) or short periods of time (seconds), which are often caused by multipath effects. Since the 1950s, many scholars have conducted research on large-scale and small-scale properties of channels in various scenarios and established a series of classical channel models, which have made great contributions to the booming communication industry today. The research on radio wave propagation reached its peak between the 1970s and the 1990s, when a large number of basic theories as well as propagation models were established. Nowadays, the research on radio wave propagation is at a stable stage of development. In recent years, research on radio wave propagation has shown three trends: new methods for predicting radio wave propagation, higher transmission frequency bands and bandwidths, and new scenarios for wireless communication applications.

The four mechanisms of propagation, that is, direct, reflection, diffraction, and scattering, are the beginning of human knowledge of radio wave propagation and the basis of wireless channel modeling. The propagation mechanism of radio waves refers to the propagation law of radio waves in the environment. For radio waves in the UHF (UHF, 300~3,000 MHz) band used in land mobile communication systems, the main propagation mechanisms are free space propagation, reflection, diffraction, and scattering. Other propagation mechanisms include transmission and absorption, waveguide, and atmospheric effects. The simplest case of radio wave propagation theory is free space propagation; i.e., a transmitting antenna and a receiving antenna exist in free space. In a more practical situation, there are also some insulating and conductive obstacles in the environment, also known as the Interacting Object (IO). If these IOs have a smooth surface, electromagnetic waves will be reflected, while another part of the energy is propagated through the interaction body. If the surface of IO is rough, the electromagnetic waves will be scattered. Eventually, electromagnetic waves will diffract at the edge of the interactor. In the following sections, the above effects will be explained in detail.

1.1.1 Free Space Attenuation

Free space is an ideal, uniform, and isotropic media space. Electromagnetic wave propagation in the media, no reflection, refraction, scattering, and absorption phenomenon, only the existence of electromagnetic wave energy diffusion caused by the propagation loss. It is worth noting that free space attenuation is not the real sense of loss, because the energy carried by electromagnetic waves is not reduced. The energy received by the receiver is reduced because the electromagnetic wave density per unit area is reduced at the receiving point after propagation over a certain distance. How to determine the "unobstructed" between the transmitter and receiver is the key to understanding the mechanism of free space propagation mechanisms. Although ray tracing reduces the propagation of electromagnetic waves to an infinitely narrow ray, in fact this "infinitely narrow" ray also has a width (wavelength is not zero), and this "width" can be described by the first Fresnel zone. The main energy of the linearly propagating waves is concentrated in the ellipsoidal body with the transmitter and receiver as the focal point. Only when there is no obstruction within this ellipsoid (first Fresnel zone), the radio wave propagation can be considered as conforming to the free space propagation mechanism. Otherwise, even if a line-of-sight distance exists between the transmitting and receiving stations and there is blocking within the first Fresnel zone, the radio wave propagation still does not conform to the free-space propagation mechanism [2].

In free space, the point source radiates energy uniformly in all directions, and the received power at any distance from the transmitter can be derived from Friis' law. The law of conservation of energy shows that the integration of energy density on any closed surface around the transmitter antenna (TX) should be equal to the transmitted power. Suppose a closed surface is a sphere with the transmitter antenna as the center and the radius is d, and the radiation of the transmitter antenna is isotropic, then the energy density of the surface is $P_{TX}/(4\pi d^2)$. The receiver antenna (RX) has an "effective area" A_{RX}. It can be assumed that all the energy impinging the area is collected by the receiver antenna, so the received energy can be expressed as follows:

$$P_{RX}(d) = P_{TX} \frac{1}{4\pi d^2} A_{RX} \tag{1.1}$$

If the transmitter antenna is not isotropic, then the energy density must be multiplied by the antenna gain G_{TX} in the direction of the receiver antenna. The product of the transmitter antenna power and gain in the considered direction is also known as Equivalent Isotropically Radiated Power (EIRP). For a given power density, the effective antenna area is proportional to the power received from the antenna connectors. It can be demonstrated that the effective area of the antenna and antenna gain exists as follows:

$$G_{RX} = \frac{4\pi}{\lambda^2} A_{RX} \tag{1.2}$$

The most noteworthy aspect of equation (1.2) is that for a fixed antenna area, the antenna gain increases with frequency. This is intuitive because the directivity of an antenna is determined by its size in terms of wavelengths. Substituting equation (1.2) into equation (1.1) yields the received power PRX as a function of the free-space distance d as a variable, a relationship also known as Friis' law:

$$P_{RX}(d) = P_{TX} G_{TX} G_{RX} \left(\frac{\lambda}{4\pi d} \right)^2 \tag{1.3}$$

In equation (1.3), $[\lambda / (4\pi d)]^2$ is known as the free space loss factor. The Free Space Attenuation has been used for a long time as a reference for the measurement and study of the propagation loss characteristics of electric waves. Since the propagation mechanism of electromagnetic waves in the atmosphere is approximately the same as that in the perfect state of free space, the direct propagation of electromagnetic waves in the atmosphere can be simplified to the propagation of direct radio waves in free space. The propagation model in free space is often used to predict the strength of the received signal when the path between the receiver and the transmitter is a completely unobstructed line-of-sight path.

1.1.2 Reflection and Transmission

The study of reflection and transmission mechanisms has benefited from the contributions of physicists in the fields of acoustics, geometrical optics, and electric wave propagation [3]. Radio waves are partially reflected and partially pass at the junction of media with different properties. If a plane wave is incident on the surface of a perfect dielectric, part of the energy enters the second media and part of the energy is reflected back to the first media without energy loss. If the second media is a perfect reflector, all the incident energy is reflected back to the first media with no energy loss. The strength of the reflected and transmitted waves depends on the Fresnel reflection coefficient Γ. The reflection coefficient is a function of the material properties and is related to polarization, angle of incidence, and frequency of the propagating wave.

In general, electromagnetic waves are polarized, which means there are instantaneous electric field components in orthogonal directions in space. A polarized wave can be mathematically expressed as the sum of two spatially orthogonal components, such as parallel and vertical, and left-hand and right-hand circularly polarized components. For a certain polarization, the reflected field can be calculated by superposition. The electric field strength of the reflected wave can be related to the incident wave in the original media by the reflection coefficient R. The magnitude of R is determined by the way the wave is polarized, the angle of incidence, and the frequency of the propagating wave, and the reflection coefficient is defined as the ratio of the magnitude of reflected E-field to the magnitude of reflected E-field given by

$$R = \frac{E_r}{E_I} = |R| e^{-j\phi} \tag{1.4}$$

where $|R|$ is the amplitude ratio between the reflected and incident E-field strengths at the reflection point, and is the phase shift of the reflected wave with respect to the incident wave.

Because of superposition, only two orthogonal polarizations need to be considered to solve the general reflection problems. At the boundary of two dielectrics, the reflection coefficients in the case of parallel and vertical polarization are given by [3]:

$$R_{\parallel} = \frac{\eta_2 \sin\theta_t - \eta_1 \sin\theta_i}{\eta_2 \sin\theta_t + \eta_1 \sin\theta_i} \tag{1.5}$$

$$R_{\perp} = \frac{\eta_2 \sin\theta_i - \eta_1 \sin\theta_t}{\eta_2 \sin\theta_i + \eta_1 \sin\theta_t} \tag{1.6}$$

where η_i is the intrinsic impedance of the ith medium, given by $\sqrt{(\zeta_i/\varepsilon_i)}$, ζ_i, and ε_i, are permeability and permittivity of the ith medium, respectively, and θ_i and θ_t are the wave incidence and refraction of electromagnetic waves, respectively.

For the ground reflected wave, if the magnitude of direct E-field strength is E_d, then the magnitude of reflected E-field strength is given by $E_d|R|e^{-j(\phi+\Delta\phi)}$, $\Delta\phi$ denoting the phase change of the incident wave. The total received field strength is the superposition of the two path waves, which is given by

$$E_t = E_d(1 + |R|e^{-j(\phi+\Delta\phi)}) \tag{1.7}$$

Equation (1.7) shows that the synthetic field of direct and ground reflected waves varies with the reflection coefficient and path difference. When φ and $\Delta\varphi$ are in phase, E_t is larger than E_d and the signal is strengthened, while, when they are opposite, E_t is smaller than E_d and the signal is weakened or canceled, and this perturbation of the signal level is fading. The reflection coefficient, at perfect ground reflection, can be shown [3] that the received signal power at a distance d between the transmitter and the receiver (T-R) is

$$P_r = P_t G_t G_r \frac{h_t^2 h_r^2}{d^4} \tag{1.8}$$

In equation (1.8), d is the distance between T-R, and h_t and h_r are heights of transmitting and receiving antennas, respectively.

Transmission refers to the incident wave entering and passing through an interacting body. Transmission is very important for light propagation into a building. If the base station is outside the building, or in another room, then the wave has to penetrate a wall (insulation) before it reaches the receiver.

1.1.3 Diffraction

The reflection phenomenon is based on the assumption that the reflecting surface is much larger than the wavelength. This assumption also limits the application of the reflection mechanism in the prediction of radio wave propagation. When radio waves encounter large obstacles in the propagation path, they will diffract the obstacles and propagate forward, and this phenomenon is called diffraction. Ultra-short wave and microwave have higher frequency, shorter wavelength, and weak bypass ability. Their signal strength behind tall buildings is small, and easy to form the so-called "shadowed region." Shadowed region signal quality is affected by the height of the building, the receiving antenna, and the distance between the building and the frequency. Diffraction makes the radio signal propagation around the curve surface of the earth be able to propagate behind the obstructions. Although the received field strength attenuates rapidly when a receiver moves into the shadowed region, the diffraction field is still present and has sufficient strength to produce a useful signal. The introduction of Huygens' principle in 1678 brought diffraction into focus, and this greatly facilitated the study of electric wave propagation. Huygens' principle showed that all points on the wave front can be used as point sources for generating secondary waves, which combine to produce a new wave front in the propagation direction. Diffraction is caused by the propagation of secondary waves into the shadowed region, where the field strength of a diffracted wave is the vector sum of all secondary waves around the obstacle. In mobile communication systems, diffraction loss occurs from the blockage of secondary waves such that only a portion of the energy is diffracted around an obstacle. In other words, an obstacle causes a blockage of energy from some Fresnel Zone, and the received energy is the vector sum of the energy contributed by the non-blocking Fresnel Zone according to the geometry of the obstruction. In general, if the obstacle does not block the first Fresnel Zone, the diffraction loss is minimal and the diffraction effect is negligible.

In 1962, Keller J B proposed the Geometrical theory of diffraction [4], which made the exact calculation of the diffraction loss possible, but the complexity of the theory prevented its application in wireless communication. The uniform geometrical theory of diffraction for an edge simplified the geometric bypass theory [5], but it still cannot avoid numerical calculations. The Knife-edge Diffraction model [6], on the other hand, is the most commonly used diffracted model due to its simplicity. Its knife-edge diffracted wave is given by

$$\frac{E_d}{E_0} = F(v) = \frac{(1+j)}{2} \int_v^\infty e^{\frac{-j\pi t^2}{2}} dt \tag{1.9}$$

where E_0 is the free space field strength and v is the Fresnel- Kirchhoff diffraction parameter, defined in a graph. The approximate loss due to the presence of a knife edge is given by [3]:

$$PL_D = \begin{cases} 0, \zeta > 1 \\ 20\log(0.5 + 0.62\zeta), 0 \le \zeta \le 1 \\ 20\log(0.5e^{0.95\zeta}), -1 \le \zeta < 0 \\ 20\log\left[0.4 - \sqrt{0.1184 - (0.1\zeta + 0.38)^2}\right], -2.4 \le \zeta < -1 \\ 20\log\left(-\frac{0.225}{\zeta}\right), \zeta < -2.4 \end{cases} \tag{1.10}$$

Subsequently, Bullington's proposed method of replacing a series of obstructions with a single equivalent obstruction [7] greatly simplified the calculations and extended the applicability of the Knife-edge Diffraction Model. However, for multi-edged obstacles, the mathematical treatment becomes a formidable mathematical problem.

Lee gave an approximate solution of the Knife-edge Diffraction Model in ref. [8], which not only facilitates the use of the model but also introduces a certain prediction error. The study of diffraction has greatly improved the system of wave propagation prediction and made more accurate deterministic modeling possible.

It is worth mentioning that the Fresnel zone theory plays a great role in the calculation of the diffraction loss. In the calculation of the edge loss, the Fresnel-Kirchoff diffraction parameter v is an independent variable in equation (1.4) which is given by

$$v = h\sqrt{\frac{2(d_1 + d_2)}{\lambda d_1 d_2}} = \alpha\sqrt{\frac{2d_1 d_2}{\lambda(d_1 + d_2)}} \tag{1.11}$$

The Fresnel radius determines which obstacles in the region will have a greater impact on the propagation of the waves, and its expression is

$$r_n = \sqrt{\frac{n\lambda d_1 d_2}{d_1 + d_2}} \tag{1.12}$$

In general, when an obstruction does not block the first Fresnel zone, the effect of diffraction can be neglected. In fact, the role of the Fresnel zone is not only reflected in the diffraction loss, but it actually determines the dissipation of radio wave energy in the process of radio wave propagation; when the Fresnel zone within the obstacle situation does not change, the energy of radio waves with the distance of the dissipation rate also remains unchanged. Numerical methods to calculate the dissipation of radio wave energy by obstacles within the Fresnel region enable the prediction of radio wave propagation in arbitrary scenarios, but such numerical calculations are often very complex, making the application of this method limited. Nevertheless, there are still a number of scholars working on this problem, such as Sweeney D G in ref. [9], who studied the effect of raindrops in the Fresnel region on radio wave propagation, and Tatarskii V I in ref. [10], who studied the effect on radio wave propagation when arbitrary obstacles are present in the Fresnel region. However, all mentioned works are based on simplifying assumptions, which makes the practicality of the results suffer.

1.1.4 Scattering

In many practical situations, scattering is common due to the large number of objects with scales comparable to the communication wavelength and the roughness of the object surfaces in the environment, which also results in stronger received signals than predicted by the diffraction and reflection models alone for actual measurements. Scattering occurs when there are objects smaller than the wavelength in the medium and the number of blocking bodies per unit volume is very large. Scattering is induced by rough surfaces, small objects, or other irregular objects such as leaves, street signs, and lampposts. Initial studies of scattering focused on the ionosphere and troposphere. Subsequently, the effects of scattering began to be considered in terrestrial mobile communication systems, and surface roughness is often tested using the Rayleigh criterion [11]. For the given angle of incidence ν, the critical height for defining the surface flatness is given by

$$h_c = \frac{\lambda}{8\sin\theta_i} \tag{1.13}$$

A surface is considered smooth if the maximum height of surface protuberances is less than h_c and the scattering in this case belongs to the reflection at the scattering point; if the height is greater than h_c, the surface is considered rough, and for rough surfaces, the surface reflection coefficient needs to be multiplied by a scattering loss factor ρ_s, to account for the diminished reflection field. Reference [12] proposed that the surface height h is a Gaussian distribution random variable with a local mean and found ρ_s is given by

$$\rho_s = \exp\left[-8\left(\frac{\pi\sigma_h\sin\theta_i}{\lambda}\right)^2\right] \tag{1.14}$$

where σ_h is the standard deviation of the surface height about the mean surface height. The scattering loss factor derived in ref. [12] was modified in ref. [13] to give better agreement with the measured results and is given by

$$\rho_s = \exp\left[-8\left(\frac{\pi\sigma_h\sin\theta_i}{\lambda}\right)^2\right] I_0\left[8\left(\frac{\pi\sigma_h\sin\theta_i}{\lambda}\right)^2\right] \tag{1.15}$$

where I_0 is the Bessel function of the first kind and zero order. The scattered field strength of a rough surface can be solved using a modified reflection coefficient given as $\rho_s R$. However, the effect of scattering is often neglected in deterministic modeling of certain scenarios due to the large attenuation experienced by scattered waves.

The four propagation mechanisms of direct, reflection, diffraction, and scattering are the beginning of human understanding of radio wave propagation and basis for the study of wireless channels. The next section will further introduce the basic properties of the wireless channel based on the basic principles of radio wave propagation.

1.2 WIRELESS CHANNELS

Wireless propagation channel is the physical medium that connects transmitter side and receiver side in a wireless communication system. The characteristics of the propagation channel directly determine the performance of wireless communication system, for example, the upper limit of the transmission capacity of the communication system, the BER of the signal at the receiving end, and the throughput of the system. An in-depth study of the wave propagation phenomenon and channel characteristics is a prerequisite for achieving optimal system design. Unlike wired channels, which are fixed and predictable, wireless channels are extremely random and are constantly affected by noise and interference as well as obstacles in the environment and their own random motion. The transmission of signals from the transmitting antenna to the receiving antenna undergoes various complex propagation paths, including direct, reflected, diffracted, scattered paths, and random combinations of these paths, which makes it important to adopt a reasonable method to accurately characterize the wireless propagation channel.

The mobile communication user's location is not fixed resulting in more unpredictable wireless channel characteristics in mobile scenarios [14]. First, the working environment of mobile communication is very complex, especially in the mobile scenario, where wireless signal will disappear with the increase of propagation distance, and will be obscured by the terrain, buildings, and the "shadow effect." The signal is reflected from multiple points and will reach the receiving location from multiple paths, and this multipath signal. The amplitude, phase, and arrival time of such multipath signals are different, and their superposition will produce level fast fading and time dispersion; secondly, communication in mobile scenes is often carried out in fast movement, which not only causes the Doppler shift and random frequency modulation, but also makes the wave propagation characteristics undergo rapid random fluctuations. Therefore, it can be considered that the wireless channels in mobile scenarios are affected by changes in time, environment and other external factors at the same time.

Regarding the characteristics of wireless channels, they can be divided into large-scale fading and small-scale fading. Large-scale fading focuses on the variation pattern of the signal in a large range (time and distance), and focuses on describing the variation of signal path loss and the characteristics of shadowing in different scenarios. The small-scale fading studies the fast fading of the signal in a short time and at a short distance. Wireless channels can also be described by narrowband and wideband communications. Narrowband communication focuses on the fluctuation of the signal amplitude, while wideband characteristics are studied for the time delay of the signal. Regardless of the division, the aim of this study is to build accurate channel models applicable to different scenarios. The modeling theory of channels is also closely integrated with the theory of radio wave propagation. In the next section, some basic properties of wireless channels are highlighted in the context of radio wave propagation mechanisms.

1.2.1 Propagation Loss and Fading

Signal propagation in free space generates loss and fading. The path loss in free space has been described in Section 1.1. This section focuses on the small-scale fading caused by path loss, shadowing, and multipath.

1.2.1.1 Path Loss

The path loss is caused by the radiation dispersion of the transmit power and the propagation characteristics of the channel together. In the path loss model it is generally considered that for the same transmitting and receiving distance, the path loss is also the same. Shadowing is caused by obstacles between the transmitter and receiver, which attenuate the signal power by absorption, reflection, scattering, and diffraction, and even block the signal in severe cases. In general, path loss causes changes in received power over long distances (10–1,000 m), while shadowing causes changes in power over obstacle scale distances (outdoor environment is 10–100 m, indoor is smaller), and both cause power changes over relatively large distances, so it is called the large scale propagation effect.

The use of path loss models allows for practical link budget design, and the commonly used path loss models can be divided into two main categories, ray tracing models, and empirical models [15]. Ray tracing models are two-path models and ten-path models. The two-path model is used in the case where a single ground reflected wave plays a dominant role in the multipath effect. The ten-path model includes various primary, secondary, and tertiary reflection signals, specifically direct paths, ground reflection paths, primary wall reflection paths, secondary wall reflection paths, wall-ground reflection paths, and ground-wall reflection paths. The calculation of path loss in all ray tracing models is performed with the transmitter and receiver positions fixed. The local average received power in the vicinity of the receiver position can be found by summing the squared amplitudes of all paths in ray tracing.

Most mobile communication systems operate in complex propagation environments that are difficult to model accurately with free-space path loss or ray-tracing methods. Over the years, many more empirical models based on measured data have been proposed for different environments, including urban macrocells, urban microcells, and models for indoor environments. The models developed are not limited to that particular environment where the data were obtained, but are to be widely applied to other general propagation environments. The Okumura model is one of the most commonly used models for signal prediction in urban macrocells, based on a large number of channel measurements in Tokyo, Japan, and gives the median loss with respect to free space propagation under irregular terrain conditions in the form of a series of curves. The Okumura model fits the path loss data represented by curves in the Okumura model to an empirical equation, and it doesn't need to determine the relevant parameters by checking the empirical curves, which greatly simplifies the calculation of path loss.

1.2.1.2 Shadowing

Shadowing is caused by blockage by other obstacles such as buildings or atmospheric refraction in the propagation path, and is a relatively flat fluctuation superimposed on the median signal. Shadowing can also be referred to as mesoscale fading, but most scholars categorize path loss and shadowing as large-scale fading. A large number of literature studies show that shadowing obeys a zero-mean log-normal distribution [3,15–18] and is calculated as follows:

$$f(m) = \frac{1}{\sqrt{2\pi\sigma^2}m}\exp[\frac{-(\ln m - \mu)^2}{2\sigma^2}] \qquad (1.16)$$

where μ is the mean value and σ is the standard deviation. However, there are exceptions: ref. [19] suggested that in the microcellular line-of-sight environment, shadowing is formed by a few stronger multipaths, and the Nakagami distribution is more consistent with the shadowing distribution of the measured data in this environment; ref. [20] suggested that the Gamma distribution is more consistent with the shadowing distribution from the theoretical analysis and fitting results of the measured data. In addition, the authors pointed out that Gamma distribution and the log-normal distribution are approximately consistent in the case of small shadowing variance.

For shadowing, the most important parameter is the standard deviation (or variance), which can reflect the large-scale fading characteristics of the propagation environment, and a large number of literature studies give the values of the variance of shadowing for different scenes and different systems [3,15]. The standard deviation of shadowing has also been further investigated by scholars such as Erceg V in ref. [21] who expressed shadowing as a random variable and found that the standard deviation of shadowing obeys a normal distribution as follows:

$$s = y\left(\mu_\sigma + z\sigma_\sigma\right) \tag{1.17}$$

However, more scholars are still stuck on the measurement of shadowing variance rather than modeling. The conclusions in ref. [21] have also been verified by few scholars at a later stage.

In addition, the correlation of shadowing has attracted the attention of many scholars. For single-antenna systems, shadowing correlation refers to the correlation of shadowing between two receiving locations at different distances from the base station. In 1991, Gudmundson [22] first investigated shadowing correlation and proposed an exponential model of shadowing correlation, which is given by

$$\begin{cases} R_A(k) = \sigma^2 \alpha^{|k|} \\ \alpha = \varepsilon_D^{vT/D} \end{cases} \tag{1.18}$$

where σ is the standard deviation of shadowing and ε_D represents the correlation coefficient between two reception points separated by a distance of D m. The model has been widely used and incorporated into standards [23,24]. The model is well adapted in suburban areas, but less well adapted in urban areas. The commonly used equivalence formula for this model is expressed as follows:

$$\rho(d) = \exp(-|d|/d_{cor}) \tag{1.19}$$

where d_{cor} is the correlation distance. At d_{cor} the correlation coefficient decreases to e^{-1}; for example, empirical measurements indicate that the correlation distance for high-speed railroad environments is mostly 120 m [23]. Reference [25] proposed an improvement to Gudmundson's simple exponential model, pointing out that when this model is extended to a continuous space, inconsistencies with the horizontal crossover theory arise. Therefore, a modified model of shadowing correlation was proposed to avoid the problem of non-convergence of the function by finding the second-order partial derivatives of this function. However, the improved model is less used because of the more complicated expressions. Today, the most popular model is still the single-exponential model shown in equation (1.19), followed by the double-exponential model mentioned in ref. [26], whose expression is

$$\rho(d) = \alpha \exp(-|d|/D_1) + (1-\alpha)\exp(-|d|/D_2) \tag{1.20}$$

where, $0 \le \alpha \le 1$, D_1 and D_2 are the short and long decorrelation distances of two components. The comparison of the exponential model and the double exponential model can be concluded that the exponential model has a higher compliance when the correlation coefficient is large, while the compliance is lower when the correlation coefficient is close to 0. This is not the case for the double exponential model.

1.2.1.3 Small-Scale Fading

Small-scale fading is used to describe the rapid fluctuations of the amplitude, phase, or multipath delays of radio signals after propagation over a short time or short distance (less than or equal to half a wavelength), so that the effect of large-scale path loss may be ignored. This fading is caused by interference between two or more versions of the transmitted signal which arrive at the receiver at slightly different times. These waves are called multipath waves, the receiver antenna combine them into a signal vary widely in amplitude and phase, depending on the distribution of the intensity, the relative propagation time of waves and the bandwidth of the transmitted signal. The fading caused by multipath is called small-scale fading. The small-scale fading can be described from both spatial and time orientations. From the spatial point of view, the amplitude of the received signal decays with distance variation along the direction of movement of the mobile station. Among them, the multipath effect caused by local reflectors shows a faster amplitude change, and its local mean value is an undulating curve with increasing distance, reflecting the fading caused by topographic undulations as well as the spatial diffusion loss. In terms of time, the length of each path is different, and the signal arrival time is different. Thus, if a pulse signal is sent from the base station, the received signal contains not only the pulse, but also its individual time delay signals. This phenomenon of expanding the width of the pulse in the received signal due to multipath effect is called time dispersion. The multipath effect of the wireless channel leads to the generation of small-scale fading, and the three most important effects are:

- Rapid changes in signal strength over a small travel distance or time interval;

- Random frequency modulation due to the varying Doppler shift on different multipath signals;

- Time dispersion effect caused by multipath propagation delay

In urban areas, fading occurs because mobile antenna is much lower than the surrounding buildings; thus, there is no line-of-sight propagation from the mobile station to the base station.

Even if such a line-of-sight propagation path exists, multipath still occurs due to reflections from the ground and surrounding buildings. Incident waves arrive in different propagation directions with different propagation time delays. The signal received by the mobile station at any point in space may consist of many plane waves, having randomly distributed amplitude, phase, and angle of incidence. These multipath components combine vectorially at the receiver antenna, causing the received signal to fad or distort. Even if the mobile receiver is stationary, the received signal will also fade due to the movement of objects in the environment in radio channel.

If the object in the wireless channel is static and the motion is generated only by the mobile station, the fading is only related to a spatial phenomenon. In this case, the spatial variations of the resulting signal are seen as temporal variations by the receiver as it moves through multipath field. Under the influence of multipath waves at different points in space, a receiver moving at high speed can pass through several fades in a very short time. In a more severe case, the receiver may stay at a specific location with a large fading. It is very difficult to maintain a good communication state, although the field model may be disturbed by passing pedestrians or vehicles, thereby diminishing the received signal to retain in a deep null for a long time. Antenna spatial diversity can prevent deep fading null. When the receiver moves within a few meters, each multipath wave undergoes an apparent shift in frequency due to the typical rapid changes in the receiver signal caused by small-scale fading and the relative motion of the mobile station and the base station. The frequency shift of the receiver signal caused by the movement is called the Doppler shift. It is related to the velocity and direction of motion of the mobile station as well as the incidence angle of the receiver multipath wave.

Many physical factors in the radio propagation channel influence small-scale fading, including the following [3]:

1. Multipath propagation

    The presence of scatterers and objects in the channel creates a constantly changing environment that dissipates signal energy, resulting in variations in signal amplitude, phase, and time. These factors result in multiple radio waves that are timely and spatially distinct from each other when the transmitting wave reaches the receiver. The random phase and amplitude of different multipath components cause signal strength fluctuations, resulting in small-scale fading and signal distortion. Multipath propagation often lengthens the time required for the baseband portion of the signal to reach the receiver which can cause signal smearing due to intersymbol interference.

2. Speed of the mobile

    The relative motion between the base station and the mobile station causes random frequency modulation due to the Doppler shift on each of the multipath components. Whether the Doppler shift is positive or negative depends on whether the mobile receiver is moving toward or behind the base station. For the motion speed of the environmental object, if the object in the wireless channel is in motion, it will cause a time-varying Doppler shift. If the environmental object moves at a speed greater than that of the mobile station, then this motion will play a decisive role in the small-scale fading; otherwise, the effect of the motion speed of the mobile station can be considered alone and the effect of the motion speed of the environmental object can be ignored.

3. Transmission bandwidth of the signal

If the transmission bandwidth of the signal is much larger than the multipath channel bandwidth, the received signal will be distorted, but the local receiver signal strength will not fade much (i.e., small-scale fading will not be significant). As will be shown, the channel bandwidth can be quantified by the coherence bandwidth, which is a measure of the maximum frequency difference and is related to the specific multipath structure of the channel. Within this range, the amplitudes of different signals remain strongly correlated. If the transmitted signal bandwidth is narrower than the channel bandwidth, the signal amplitude changes rapidly, but the signal will not be distorted in time. Therefore, the amplitude of small-scale signals and the likelihood of signal smearing after short-range transmission are related to the specific amplitude of the multipath channel, the time delay, and the bandwidth of the transmitted signal.

1.2.2 Frequency Selective Fading

Time-varying multipath channels can be characterized in frequency domain by making a Fourier transform of time-varying channel impact response $h(\tau,t)$ with respect to τ. The result of the Fourier transform is a stochastic process:

$$H(f,t) = \int_{-\infty}^{\infty} h(\tau,t)e^{-j2\pi f\tau}\,d\tau \tag{1.21}$$

Since $h(\tau,t)$ is a zero-mean complex Gaussian stochastic process with t as the time covariate, the Fourier transform of equation (1.21) is a summation of the zero-mean complex Gaussian process, so $H(f,t)$ is also a zero-mean complex Gaussian process. Its properties are completely determined by the autocorrelation function. And since $h(\tau,t)$ is a wide sense stationary process, $H(f,t)$ is also a wide sense stationary process. In this way, the autocorrelation function of equation (1.21) can be expressed as

$$A_H\left(f_1,f_2,;\Delta t\right) = E\left[H^*\left(f_1;t\right)H\left(f_2;t+\Delta t\right)\right] \tag{1.22}$$

This can be simplified to the following form:

$$A_H\left(f_1,f_2,;\Delta t\right) = E\left[\int_{-\infty}^{\infty} h^*(\tau_1;t)e^{-j2\pi f_1\tau_1}\,d\tau_1 \int_{-\infty}^{\infty} h(\tau_2;t+\Delta t)e^{-j2\pi f_2\tau_2}\,d\tau_2\right]$$

$$= \int_{-\infty}^{\infty}\int_{-\infty}^{\infty} E\left[h^*(\tau_1;t)h(\tau_2;t+\Delta t)\right]e^{-j2\pi f_1\tau_1}e^{-j2\pi f_2\tau_2}\,d\tau_1\,d\tau_2 \tag{1.23}$$

$$= \int_{-\infty}^{\infty} A_h\left(\tau,\Delta t\right)e^{-j2\pi(f_1-f_2)\tau}\,d\tau = A_H\left(\Delta f,\Delta t\right)$$

where, Δf is given by $\Delta f = f_2 - f_1$ in equation (1.23), and the third equation comes from the wide sense stationary process and uncorrelated scatterers assumption of $h(\tau,t)$ (see Section 1.2.4). Thus, the autocorrelation function of $H(f,t)$ with respect to frequency is only related to Δf. $A_H(\Delta f, \Delta t)$ is measured by sending a pair of sinusoidal waves with a carrier frequency difference of Δf and then finding the correlation value of these two sine waves at the receiver after a time interval of Δt.

If $A_H(\Delta f) \equiv A_H(\Delta f, 0)$ is defined, then from equation (1.23) we have

$$A_H(\Delta f) = \int_{-\infty}^{\infty} A_h(\tau) e^{-j2\pi \Delta f \tau} \, d\tau \qquad (1.24)$$

That is, $A_H(\Delta f)$ is the Fourier transform of the power delay profile, and since $A_H(\Delta f) = E[H^*(f,t)H(f+\Delta f,t)]$ is an autocorrelation function, the channel responses at frequencies separated by Δf are approximately independent if $A_H(\Delta f) \approx 0$. If there exists some B_c such that for all $\Delta f > B_c$ there is $A_H(\Delta f) \approx 0$, then B_c is called the Coherence Bandwidth of the channel. From the Fourier relationship between $A_h(\Delta \tau)$和$A_H(\Delta f)$, if we have $A_H(\Delta f) \approx 0$ for $\tau > T$, we have $A_H(\Delta f) \approx 0$ for $\Delta f > 1/T$. Therefore, the minimum frequency interval that can make the channel response approximately independent is $B_c \approx 1/T$, and T is generally taken as $A_h(\Delta \tau)$ of the root-mean-square delay expansion σ_{Tm}. A more general approximation is $B_c \approx k/\sigma_{Tm}$, with k depending on the shape of $A_h(\Delta \tau)$ and exactly how the coherence bandwidth is defined. It was noted that the frequency range for channel correlation greater than 0.9 is $B_c \approx 0.02/\sigma_{Tm}$ and for channel correlation greater than 0.5 is $B_c \approx 0.2/\sigma_{Tm}$ [15].

In general, if a narrowband signal with bandwidth $B \ll Bc$, then the fading within the signal bandwidth is highly correlated, i.e., the fading is approximately equal over the entire bandwidth, which is usually referred to as flat fading. On the other hand, if the signal bandwidth $B \gg Bc$, then the channel amplitude values at the two frequency points separated by more than the coherent bandwidth are approximately independent. At this time, the channel amplitude varies greatly within the signal bandwidth, and such a fading is called frequency selective fading. When the situation at $B \approx B_c$ is between flat fading and frequency selective fading, the signal bandwidth B of linear modulation is inversely proportional to the code-element interval T_S, and flat fading corresponds to $T_S \approx 1/B \gg 1/B_c = \sigma_{Tm}$, when intersymbol interference can be ignored; frequency selective fading corresponds to $T_S \approx 1/B \ll 1/B_c = \sigma_{Tm}$, when there will be serious intersymbol interference. Although multi-carrier modulation and spread spectrum techniques can reduce intersymbol interference, they are wideband signal approaches, and thus frequency selective fading also occurs throughout the signal bandwidth, which will degrade the system performance.

The power delay profile $A_h(\tau)$ and its Fourier transform $A_H(\Delta f)$ are given in Figure 1.1. Both a narrowband signal with a bandwidth much smaller than B_c and a broadband signal with a bandwidth much larger than B_c are shown in the $A_H(\Delta f)$ plot. Within the bandwidth of the narrowband signal, the autocorrelation function $A_H(\Delta f)$ is flat and the signal will experience flat fading and intersymbol interference can be neglected. Within the bandwidth

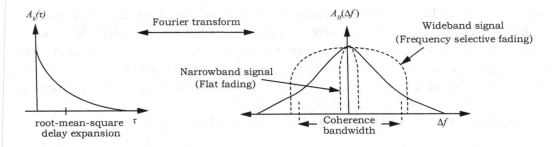

FIGURE 1.1 Example of Power delay profile, root-mean-square delay expansion, and coherence bandwidth [15].

of the wideband signal, the autocorrelation function $A_H(\Delta f)$ gradually decreases to zero, indicating that the fading in different parts of the signal bandwidth is independent of each other, and therefore is frequency selective fading, and the linearly modulated signal transmitted over this channel will experience severe intersymbol interference.

1.2.3 Doppler Effect

The Doppler effect is caused by the relative motion between the mobile station and the base station, or by the motion of other objects in the radio environment. Suppose that the mobile station receives a signal from a distant source S at a constant velocity v, while moving on a path of length d with endpoints X and Y, as shown in Figure 1.2. At this point, the radio waves from the source S, at the X and Y points, are received by the mobile station when the path difference is given as $\Delta l = d \cos \theta = v \, \Delta t \cos \theta$. The time required for the mobile station to move from X to Y is Δt; θ is the angle with the incident wave at X and Y. Since the source end is far away, θ at X and Y can be assumed to be the same. Therefore, the value of the phase change of the received signal caused by the distance difference is

$$\Delta \varphi = \frac{2\pi \Delta l}{\lambda} = \frac{2\pi v \Delta t}{\lambda} \cos \theta \tag{1.25}$$

Doppler shift f_d is given by

$$f_d = \frac{1}{2\pi} \cdot \frac{\Delta \varphi}{\Delta t} = \frac{v}{\lambda} \cdot \cos \theta \tag{1.26}$$

From equation (1.26), it can be seen that the Doppler shift is related to the movement speed of the mobile station, the direction of movement of the mobile station, and the angle between the incident directions of the radio waves. If the mobile station moves toward the incident direction, the Doppler shift is positive (i.e., the received frequency increases); if the mobile station moves backward toward the incident direction, the Doppler shift is negative (i.e., the received frequency decreases). The signal propagates in different directions, and its multipath component causes Doppler spread of the receiver signal.

If the mobile station moves, the multipath components arrive at the mobile station from different directions and cause different frequency shifts, which results in an expansion of the

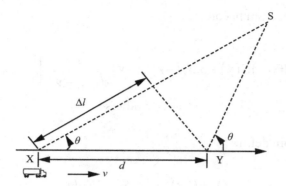

FIGURE 1.2   Illustration of the Doppler effect.

received spectrum. It is assumed that the transmitted signal is sinusoidal (i.e., narrowband case). When a wave arrives from a single direction only, the expression for the Doppler shift can be rewritten. γ represents the angle between the velocity vector v of the moving station and the direction of the wave at the moving station. As shown in equation (1.27), the Doppler effect causes the received frequency f to be shifted by v, so that the received frequency is

$$f = f_c \left[1 - \frac{v}{c_0} \cos\gamma \right] = f_c - v \tag{1.27}$$

where $v = |v|$. Apparently, the frequency offset depends on the direction of the wave and must be within the range $f_c - v_{max} \sim f_c + v_{max}$, where $v_{max} = f_c v/c$.

If there are multiple multipath components, it is necessary to know the distribution of the incident wave power as a function of the variation of γ. Since the statistical distribution of the received signal is of interest, the probability density function of the received power $\mathrm{pdf}_\gamma(\gamma)$ is considered. As a slightly misused notation, the authors call it the probability density function of the incident wave. The multipath component arriving at the receiver is weighted by the antenna pattern of the mobile station. Thus, the multipath component arriving in the direction γ has to be multiplied by the mode $G(\gamma)$. In this way, the received power spectrum as a function of direction is given as follows:

$$S(\gamma) = \bar{\Omega} \left[\mathrm{pdf}_\gamma(\gamma) G(\gamma) + \mathrm{pdf}_\gamma(\gamma) G(\gamma) \right] \tag{1.28}$$

where $\bar{\Omega}$ is the average power of the arrival field. In equation (1.19), the authors also found that the waves coming from γ and $-\gamma$ directions lead to the same Doppler shift, so there is no need to distinguish for the derivation of the Doppler spectrum. In the last step, it is necessary to perform the transformation of the variables, γ to v, and the Jacobian matrix can be determined as

$$\left| \frac{d\gamma}{dv} \right| = \frac{1}{\left| \dfrac{dv}{d\gamma} \right|} = \frac{1}{\left| \dfrac{v}{c_0} \sin\gamma \right|} = \frac{1}{\sqrt{\left(f_c \dfrac{v}{c_0} \right)^2 - \left(f - f_c \right)^2}} = \frac{1}{\sqrt{v_{max}^2 - v^2}} \tag{1.29}$$

Thus, the Doppler spectrum becomes:

$$S_D(v) = \begin{cases} \bar{\Omega} \left[\mathrm{pdf}_\gamma(\gamma)G(\gamma) + \mathrm{pdf}_\gamma(-\gamma)G(-\gamma)\dfrac{1}{\sqrt{v_{max}^2 - v^2}} \right], -v_{max} \le v \le v_{max} \\ \\ 0, \text{others} \end{cases} \qquad (1.30)$$

For further application, define

$$\Omega_n = (2\pi)^n \int_{-v_{max}}^{v_{max}} S_D(v) v^n \, dv \qquad (1.31)$$

where Ω_n is the nth order moment of the Doppler spectrum. The specific angular distribution and antenna pattern correspond to their specific formulas. A widely adopted model for the angular spectrum at mobile stations is that the waves are incident uniformly from all azimuthal directions and all reach the horizontal plane, thus:

$$\mathrm{pdf}_\gamma(\gamma) = \frac{1}{2\pi} \qquad (1.32)$$

This case corresponds to the situation where there is no line-of-sight connection and a large number of interacting bodies are uniformly distributed around the mobile station. Further assuming that the antenna is a vertical dipole antenna with the antenna pattern $G(\gamma)$, the Doppler spectrum is given by

$$S_D(v) = \frac{1.5\bar{\Omega}}{\pi\sqrt{v_{max}^2 - v^2}} \qquad (1.33)$$

This spectrum is the classical spectrum or Jakes spectrum, as shown in Figure 1.3. It has a very typical "U" shape and there are (integral) singularities at the minimum and maximum Doppler frequencies $v = \pm v_{max} = \pm f_c v/c$. These singularities correspond to the direction of movement of the receiver or its inverse. It is worth noting here that a uniform directional distribution leads to a highly non-uniform Doppler spectrum.

In fact, the Doppler spectrum of actual channel measurements does not show a singularity. Even if the model and assumptions are strictly valid, it takes an infinite number of measurement samples to reach a singularity. Even so, the classical Doppler spectrum is still one of the most widely used Doppler spectrum models.

1.2.4 The WSSUS Model

In wireless channel theory, there exist two commonly used assumptions for simplified channel models, namely the so-called Wide-Sense Stationary (WSS) channel and Uncorrelated Scattering (U.S.) channel [11].

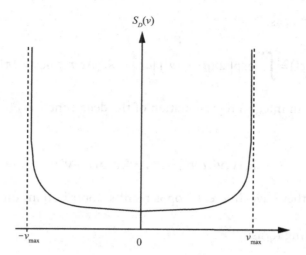

$S_D(v)$

$-v_{max}$ 0 v_{max}

FIGURE 1.3   Classical Doppler spectrum.

1.2.4.1 Wide-Sense Stationarity

The mathematical definition of Wide-Sense Stationarity is that the autocorrelation function is independent of the variables t, t', but depends on their difference $t - t'$. Consequently, the second-order amplitude statistics do not change with time. Thus, we can write equation (1.34):

$$R_h(t,t',\tau,\tau') = R_h(t,t+\Delta t,\tau,\tau') = R_h(\Delta t,\tau,\tau') \qquad (1.34)$$

Physically speaking, WSS means that the statistical properties of the channel do not vary with time. This is not to be confused with static channels, where the fading realization is not time-varying. For the simple case of the flat Rayleigh-fading channel, WSS implies that both the mean power and the Doppler spectrum do not vary with time, while the instantaneous amplitude can change.

According to the mathematical definition, WSS has to be fulfilled for any arbitrary time, t. In practice, this is not possible; as the mobile station moves over larger distances and the mean received power changes due to shadowing and variation in path loss. Rather, the typical case of WSS means that it is fulfilled over an area of about 10λ diameter. This allows to define a quasi-stationarity (related to the distance the mobile station moves) for a finite time interval, during which the statistics do not change significantly.

Doppler-variant impulse response $S(v,t)$ provides a further interpretation of the WSS. The extended function is denoted as.

$$R_s(v,v',\tau,\tau') = \int_{-\infty}^{\infty}\int_{-\infty}^{\infty} R_h(t,t+\Delta t,\tau,\tau')\exp\left[2\pi j\left(vt - v'(t+\Delta t)\right)\right]dt\,dt' \qquad (1.35)$$

which can be rewritten as:

$$R_s(v,v',\tau,\tau') = \int_{-\infty}^{\infty} \exp\left[2\pi jt(v-v')\right]dt \int_{-\infty}^{\infty} R_h(\Delta t,\tau,\tau')\exp\left[2\pi jv'\Delta t\right]d\Delta t \quad (1.36)$$

The first integral is an integral representation of the delta function $\delta(v-v')$. Thus, R_s can be factored as:

$$R_s(v,v',\tau,\tau') = P(v,v',\tau)\delta(v-v') \quad (1.37)$$

This implies that if they have different Doppler shifts, contributions undergo uncorrelated fading.

Analogously, noting R_B as:

$$R_B(v,v',f,f') = P_B(v,f,f')\delta(v-v') \quad (1.38)$$

1.2.4.2 Uncorrelated Scatterers

US is defined as follows: contributions with different time delays are uncorrelated and can be expressed mathematically as:

$$R_h(t,t',\tau,\tau') = P_h(t,t',\tau)\delta(\tau-\tau') \quad (1.39)$$

or for R_S as:

$$R_S(v,v',\tau,\tau') = P_S(v,v',\tau)\delta(\tau-\tau') \quad (1.40)$$

The US condition is fulfilled when one multipath component does not contain any phase information of another multipath component with different time delays. If the scatterer is randomly distributed in space, the phase changes in an uncorrelated way even if the mobile station moves a small distance. For the transfer function, the uncorrelated scattering condition means that R is independent of the absolute frequency, but only related to the frequency difference:

$$R_H(t,t',f,f+\Delta f) = R_H(t,t',\Delta f) \quad (1.41)$$

1.2.4.3 WSSUS Assumption

The US and the WSS assumption are duals: the US assumption defines that the different time delays of the scatterers are uncorrelated with each other, while the WSS considers that the different Doppler shifts of the scatterers are uncorrelated with each other. In addition, it can be considered that the R_H of the US is related to the frequency difference only, while the R_H of the WSS assumption is related to the time difference only. Obviously, the combination of these two assumptions is the WSSUS condition. So that the autocorrelation function has to fulfill the following conditions:

$$R_h(t, t+\Delta t, \tau, \tau') = \delta(\tau - \tau') P_h(\Delta t, \tau) \qquad (1.42)$$

$$R_H(t, t+\Delta t, f, f+\Delta f) = R_H(\Delta t, \Delta f) \qquad (1.43)$$

$$R_S(v, v', \tau, \tau') = \delta(v - v') \delta(\tau - \tau') P_S(v, \tau) \qquad (1.44)$$

$$R_B(v, v', f, f+\Delta f) = \delta(v - v') P_B(v, \Delta f) \qquad (1.45)$$

Unlike the autocorrelation functions that depend on four variables, the P-functions on the right side of the computational equation depend on only two variables. Their formal description, parameterization, and application can be greatly simplified in further derivations. Because of their importance, they are given different names. They are given as follows:

- $P_h(\Delta t, \tau)$ as delay cross power spectral density;
- $P_H(\Delta t, \Delta f)$ as time frequency correlation function;
- $P_S(v, \tau)$ as scattering function;
- $P_B(v, \Delta f)$ as Doppler cross power spectral density.

The scattering function has a special importance, if a single interaction occurs, then each differential element of the scattering function corresponds to the physically existing interacting body. From the Doppler shift the direction of arrival can be determined Direction Of Arrival (DOA) and the time delay can determine the radius of the ellipse where the scattering exists.

The WSSUS assumption is very commonly used in wireless communications, but not always fulfilled in practice. However, it is important to note that the WSSUS assumption is based on the premise that there are a large number of independent scatterers in the channel and is initially oriented towards ionospheric channels. For terrestrial wireless channels, this is often a necessary premise that is overlooked. Next, the validity of this assumption is discussed in detail [1]. These considerations may seem at first glance to be somewhat convoluted and abstract; thus, it is very necessary to understand the constraints of the WSSUS assumption.

A key issue is the definition of expectation. Mathematically, it is clear that the overall data set is used to derive the expectation. But what is the "overall data set" for a physical channel? If only one moment is considered, then the propagation scenario is completely deterministic, and the impulse response of the channel can be calculated by Maxwell's set of equations. Similarly, the measured impulse response corresponds to a well-defined set of measurement parameters (time, location, frequency range, etc.), so that the measurement data set used in the modeling actually has some underlying assumptions built into it. Most commonly, the impulse responses measured at different times are treated as a set of measurement data sets, which potentially implies that the channel is ergodic (meaning that the statistical average of the channel over a period of time is equivalent to the overall data set average), which is important for the WSSUS assumption.

The WSS condition implies that the Auto Correlation Function (ACF) must be independent of the absolute time. It is clear from the physical propagation mechanism that the average received power (actually the value of the ACF at the point where the time delay is zero) decreases as the mobile station (MS) moves away from the base station (BS). Similarly, the statistical properties of the channel change when the MS moves from the Line of Sight (LOS) state with the BS to the "shadow" area at the back of the building. Scholar Bello proposed the concept of Quasi-Stationarity. The Quasi-Stationarity interval is defined as the time period for which the WSS can be satisfied, and the WSS assumption cannot be satisfied for distances beyond this interval (e.g., if the MS moves beyond this interval) [27]. This concept is obviously different from the strict mathematical definition of the smooth concept, since it does not require the assumption of an infinite reference time interval. It also applies to the inscription of instantaneous stationary intervals – too large an interval and the channel is non-stationary; too small an interval and the number of valid samples is too small. This is particularly evident in indoor channels, where the average received power changes significantly over very short distances.

Further, any ergodic process is generalized stationary. Thus, once the ergological condition is used to obtain the complete data set, generalized stationary is effectively defaulted – no matter how large a time interval (which can be larger than a stationary interval) the data set is taken. This seeming contradiction can be explained by the fact that a time-series process can be averaged using a "sliding window" to observe changes in the statistical characteristics of the process. In other words, two measurements within a short time interval can give some information about the stationary. In the full data set, this information is lost – because all data samples can be considered as time-series independent.

Uncorrelated scattering (US) suggests that the fading due to different scatterers is uncorrelated. This assumption condition is applicable for most outdoor scenarios, but it is often not applicable for indoor environments, especially for empty rooms. For example, the fading due to reflection paths from interior walls is significantly correlated.

1.2.5 Multipath Clusters

The transmission medium for wireless communication is the wireless channel between the transmitter and the receiver. Signals can propagate from the transmitter to the receiver through different propagation paths. In some cases, a line-of-sight (LOS) propagation path exists between the transmitter and the receiver. In addition, there are different interacting bodies in the radio wave propagation environment (e.g., houses and mountains, windows, walls, etc. in outdoor environments), and the signal can reach the receiver from the transmitter by means of reflections from the surfaces of these interacting bodies or bypasses by their obstruction. These different propagation paths are called "multipath" by the authors [1]. At a higher level, the energy of multipaths in the real environment tends to reach the receiver "in groups," so that the simultaneous arrival of multipaths can be defined as a "cluster." In the study of radio wave propagation, there is no complete consensus on the definition of multipath clusters, which originated from the observation in wireless channel measurements, and the application of the multipath cluster concept can improve the performance of channel modeling. Multipath exists in the form of clusters in real channels mainly because of the discrete distribution of scatterers in the physical channel, each of which forms one or several paths through which wave energy is transmitted from the transmitter to the receiver. At the receiver, these clusters arrive at different times and from different angles, but the multipath signals within each cluster are very similar. Therefore, clusters can be observed in the time delay domain, and in the angle domain. The size of the clusters is mainly determined by the time delay/angle expansion observed by the receiver and the transmitter. Also, multipath clusters in a wireless channel can be further subdivided into several multipaths, each of which has its own propagation delay, Angle of Arrival (AOA), Angle of Departure (AOD), and power. Figure 1.4 shows the schematic multipath distribution model based on the cluster structure, where the multipath clusters can be further decomposed into some distinguishable multipath components. Cluster-based channel modeling can further abstract and simplify the variation of propagation characteristics in the channel without causing significant degradation in modeling accuracy. For most of the cluster models, the parameters of multipath clusters are extracted from the measurement data of the real environment, so the cluster models can also reflect the statistical properties of the real channel well. In addition, in order to make the model support larger bandwidth and improve the resolution of the delay domain in channel modeling, the clusters with higher energy in the channel model can be divided into several sub-clusters, which can keep the total number of multipaths in the model constant and add additional delay taps.

Each cluster is composed of multiple multipath components, so a cluster can be considered abstractly as a set of multipath components with similar parameters. Identifying a cluster starts with determining how the multipath components within each cluster are grouped together, or how similar the parameters of the multipath components are to each other. Cluster identification can be observed intuitively with the help of the time-delay-angle spectrum, or automatically using various clustering algorithms. The intuitive method of observation to make judgments about clusters is often subjective and may result in different researchers giving very different clustering results for the same set of data. In addition, when the test data are relatively large, the conducting multipath

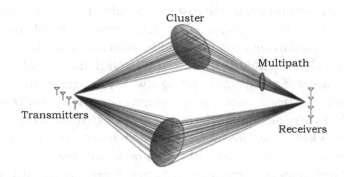

FIGURE 1.4   Schematic of the cluster-based model.

clustering is also a more laborious task [28]. Nevertheless, this method has been widely used and was the main method for the initial understanding of the multipath clustering phenomenon.

The emergence of multipath clustering has had a great impact on the development of channel modeling. Turin et al. [29] were the first to experimentally observe that radio wave energy arrives at the receiver in the form of clusters and that the arrival of clusters obeys a Poisson process. Saleh and Valenzuela built on Turin's work by stating that radio wave energy also arrives in broadband channels in the form of clusters. The concept of clusters was introduced to explain the multipath phenomenon in the time delay domain when Saleh and Valenzuela, based on Turin's work, showed that radio wave energy also arrives in the form of clusters in broadband channels [30] and each cluster is composed of several distinguishable multipath components. The cluster-based concept was extended to the azimuth-delay domain in the COST259 model [31,32], and Laurila et al. partitioned clusters in the azimuth-delay domain [33]. Chong et al. [34] proposed a statistical wideband channel model using a single directional model and the concept of subclustering. Yu et al. [35] investigated the angular extension of clusters using test data of wideband channels in an indoor environment, providing a reference for modeling the IEEE 802.11 TGn standard. Czink et al. [36] partitioned clusters for the first time in the azimuth-delay domain in a two-way channel. This is a great advancement in cluster partitioning, where clusters are easier to partition in the AOA-AOD domain than in the delay domain, and clusters are more concentrated in the AOA-AOD domain.

Multipath clustering is utilized in channel modeling, thus reducing model complexity while maintaining modeling accuracy. The earliest channel model that covers the multipath cluster structure is the SV (Saleh-Valenzuela) model. In this model, the multipath components are divided into different clusters in the time delay domain based on the measured data. In addition, a geometric stochastic channel model (GSCM), which is more suitable for MIMO channels, has been proposed to extend the multipath delay clusters in the SV model to both delay and angle dimensions. The multipath clustering phenomenon in the angle dimension has also influenced the development of the Double-Directional channel model [37]. With the help of some classical multipath parameter estimation methods, such as the Space-Alternating Generalized Expectation Maximization (SAGE) algorithm [38] and the improved SAGE algorithm ISIS [39], directional channel models based on cluster structure are established. In the past two decades, multipath clustering has been widely observed in many environments, while channel models based on cluster structure are widely used in standardized channel models, such as COST259, COST2100, 3GPP Spatial Distribution Channel Model (SCM), and WINNER model [40,41].

Although the concept of multipath clustering is widely recognized in channel modeling, establishing suitable multipath clustering algorithms remains a popular topic [42,43]. In the past, methods to carry out multipath clustering through human visual discrimination have been used for a long time. However, even though human vision can effectively identify the structure and pattern of multipaths from noise, this method is too cumbersome in the face of a large amount of measurement data. Therefore, channel modeling based on cluster results requires an automated clustering algorithm with sophisticated design.

While clustering algorithms (mostly called "clustering algorithms" in machine learning) have been a popular research topic in machine learning, the problem of channel multipath clustering is still an emerging discipline in the field of wireless communications. Since there are many parameters describing the multipath components in the actual propagation channel, including power, delay, angle, etc., and all of them have real physical meaning and different statistical properties, the biggest challenge of multipath clustering lies in how to take the effects of these parameters into account in the clustering algorithm. Some multipath clustering algorithms exist when only power and delay information are considered; however, such algorithms are only applicable in the delay domain of multipath components and cannot be used for multipath clustering of MIMO channels (covering multipath angular domain distribution characteristics).

In recent years, several algorithms for automatic clustering have been proposed, and automatic clustering algorithms provide a repeatable and automatic method for clustering, each giving its own definition of clusters. Salo et al. [44] gave a semi-automatic method for dividing clusters in the azimuth and time delay domains separately and used the concept of hierarchical clustering. By performing independent estimation of multipath clusters in different dimensions, the estimated clusters in multiple dimensions can be easily combined to improve the clustering accuracy. In ref. [36], Czink gives the K-Power-Means (KPM) clustering algorithm, and the algorithm is able to perform cluster partitioning over the AOA-AOD-delay domain. In the analysis, the effects of both horizontal and azimuth angles are considered. Czink proposed a Random Cluster Model (RCM) based on an improved COST273 model in ref. [44]. Currently, the main algorithms applicable to multipath clustering of MIMO channels considering all multipath parameters (power, delay, angle) are summarized as follows: in ref. [45], the KPM algorithm is proposed, which considers the effect of multipath power in the calculation of cluster centers and uses multipath distance to define the similarity between different multipath components. In ref. [46], the Fuzzy c-means algorithm was improved and used for multipath clustering, and it was shown that this algorithm outperforms the KPM algorithm under the condition of random initialization.

The multipath clustering problem for time-varying channels in mobile scenarios is one of the main difficulties in current multipath clustering research, because there are always rapid changes in multipath signals in time-varying channels, and their multipath similarity and clustering characteristics are difficult to be captured. To address this problem, an automatic clustering and tracking algorithm for multipath components of time-varying channels is proposed in ref. [47]. The proposed algorithm is based on the dynamic characteristics of the channel in the time domain, and thus can naturally reflect the birth and death characteristics of the multipath components. The algorithm is validated using ray tracer and SCME simulation data, and accurate clustering and tracking results are obtained. Compared with existing clustering algorithms, the algorithm proposed in ref. [47] is able to cluster and track time-varying multipath components with high accuracy and low complexity, and is described in detail next.

In most cases, the multipath components of a time-varying channel are captured and stored at discrete moments over many short time intervals $S_1, S_2, \ldots, S_{i-1}, S_i, S_{i+1}, \ldots$. Therefore, the objective is to cluster the multipath components at each time moment and to find the relationship between the clusters of adjacent time moments for cluster tracking. Unlike the existing clustering algorithms, the proposed clustering algorithm in ref. [47] consists of the following two steps: the initial clustering of the multipath components at the first time moment using the existing clustering algorithm, and the clustering of the multipath components at the next time moment by finding the time-varying relationships between the clusters at the neighboring time moments. The detailed implementation is described below.

Initialization: It is important to note that each multipath component corresponds to an independent backscattering point. These points (or multipath components) can be clustered by existing clustering algorithms, such as the standard K-means algorithm or the DBSCAN algorithm. In the process of clustering using these algorithms, the multipath power is also utilized for weighting, thus better matching the actual situation. With the initial clustering, the initial cluster center positions of the K clusters are obtained and numbered sequentially (from 1 to K).

Cluster and track for multipath components: The initialization step has completed the clustering of the multipath components. The next goal is to find the relationship between the clusters at adjacent time moments, so that the multipath components at the remaining time moments can be tracked. The process consists of the following two steps.

Step 1 First roughly define the intrinsic relationship of the clusters between adjacent time moments S_{i-1} and S_i

Assume that k is the cluster number and j is the number of the multipath component, and u_k is the cluster center of the kth cluster M_k. Consider the multipath components of the remaining time points $S_2,\ldots,S_{i-1},S_i,S_{i+1}$ as a data set Y except for the first time point S_1, and the data points $y_j \in Y$ of the current time point S_i will be clustered on the results of the previous time point S_{i-1}:

1. Calculate the multipath distance $d(y_j,u_k)$ between the data points y_j at the current time moment S_i and all cluster centers at the previous time moment S_{i-1}. In the process of calculating the multipath distance, the weight of the delay can also be changed by defining a delay variation factor, which is set to 2 in the validation process based on the simulation data to obtain a more reasonable result. Next, find the cluster in S_{i-1} with the closest distance to data point Y_i in S_i by calculating the distance M_k. Naturally, the data point y_i is attributed to the cluster M_k, and the corresponding number of the cluster M_k is assigned to y_i.

2. After all data points of S_i at the current time moment are numbered and assigned, the data of S_{i-1} at the previous time moment will be completely deleted, thus leaving the complete clustering result of S_i.

So far, only the approximate connection of clusters between adjacent time moments has been obtained. Obviously, due to the existence of time-varying multipath component births and deaths, the cluster inheritance relations are not always as described, so further research is needed.

Step 2 In this step, a new cluster in S_i at the current time moment is defined. The following two thresholds are defined based on the clustering results of the previous time moment S_{i-1}.

Threshold 1: The maximum range of each cluster. The maximum range of the kth cluster $TH_{1,k}$ is defined as the multipath distance corresponding to the multipath component that is farthest from the cluster center among all the multipath components in that cluster:

$$TH_{1,k} = \max_j \left\{ d\left(x_j, u_k\right) \right\} \tag{1.46}$$

where, $x_j \in X$ and X is the set of data at the previous time moment S_{i-1}.

Threshold 2: Minimum density of all clusters. Minimum density of all clusters TH_2 is defined as the smallest number of scatterers per unit area in all existing clusters, with the aim of distinguishing between the effective multipath component and the noise:

$$TH_2 = \min_k \left\{ \frac{\text{number}(k)}{\pi \times \left[d\left(x_j, u_k\right) \right]^2} \right\} \tag{1.47}$$

where, $\text{number}(k)$ represents the number of scatterers in the kth cluster. It is important to note that both thresholds are defined based on the results of the previous time moment S_{i-1} for cluster splitting.

Using the above two thresholds, a new cluster for the S_i at the current time moment will be defined in the following way.

First, the threshold 1 determines whether a complete new cluster can be created. If the data points do not fall into the maximum range of the corresponding cluster, a new cluster number is temporarily assigned; however, a new cluster is defined only if the number of points in the same new cluster is greater than a specific value δ; otherwise, these points are marked as noise and no corresponding number is assigned to them. Regarding the value of δ: depending on the scenario and the data format, the value of δ should be large enough to avoid that wrong new clusters are defined; however, the value should not be larger than the minimum value of the multipath components of all existing clusters, which can be obtained from the a priori information.

Secondly, to avoid noise being defined as new clusters, based on threshold 2, the new clusters defined in ② will be checked. If the density of the defined new cluster is less than TH_2, this new cluster and its corresponding number will be deleted and these points will also be marked as noise at the current moment and no corresponding number will be assigned to them.

Based on the above steps, the cluster centers of all clusters in the current moment S_i can be calculated. In order to avoid that the multipath component of low power has a similar effect on the clustering results as the multipath component of high power, the multipath power is further taken into account. Thus, the calculations of cluster cores are given as

$$u_k^{(i)} = \frac{\sum_{j \in M_k^{(i)}} P_j \cdot y_j}{\sum_{j \in M_k^{(i)}} P_j} \tag{1.48}$$

where $u_k^{(i)}$ and P_j denote the cluster center $M_k^{(i)}$ and the power of the jth multipath component, respectively, of the cluster at the current time S_i. The cluster core and cluster number information are used to determine the cluster in the next time moment S_{i+1}.

In this algorithm, a pre-existing cluster will always inherit the number of this cluster from the previous moment, while a new cluster will be assigned a new number. Thus, the temporal evolution of clusters can be tracked by identifying their numbers. Moreover, the whole tracking process can be automated without corrections. A schematic representation of the aforementioned clustering and tracking algorithm is given in Figure 1.5, and a more detailed analysis can be found in ref. [47].

Figure 1.5 shows a schematic of the clustering and tracking algorithm [47]. Suppose there are two clusters M_1 and M_2, whose cluster centers are u_1 and u_2, respectively, $\delta=3$. There are ten points at the time $y_1, \ldots y_{10}$.

Step 1 By calculating the multipath distance, find the clusters of all points at the time S_i that are closest to the time S_{i-1}, so that $y_1, y_2, y_3, y_7, y_8, y_9$ belongs to M_1, and y_4, y_5, y_6, y_{10} belongs to M_2.

Step 2 Determine whether a new cluster can be defined for the six points $y_1, y_2, y_3, y_7, y_8, y_9$ based on the threshold $TH_{1,1}$. Based on the same criteria, y_{10} is determined to

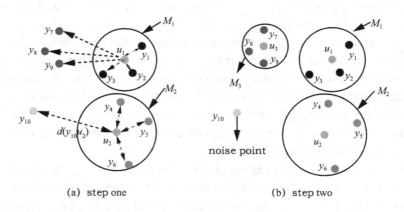

(a) step one   (b) step two

FIGURE 1.5 Schematic representation of the aforementioned clustering and tracking algorithm.

be noisy and no number will be assigned to it. Next, the new cluster formed by y_7, y_8, y_9 is determined to be valid according to the threshold TH_2, and the number M_3 is assigned to this new cluster. Finally, the cluster cores of the three clusters at the S_i moment are further calculated by using equation (1.48).

Although some progress has been made in the last decade in the study of multipath automatic clustering algorithms for wireless channels, the existing work still has the following limitations.

- Many parameters of the multipath component are not considered in the clustering algorithm. Unlike artificially generated data in machine learning, multipath signals in real scenarios are generated by the physical environment and have defined intrinsic physical characteristics. The physical laws of these multipath components should be taken into account in the clustering algorithm. For example, many measurements show that the angular distribution of multipath clusters usually obeys the Laplace distribution; however, this property is not taken into account in the design of existing clustering algorithms.

- The number of multipath clusters in existing algorithms is often required as known information to be input into the clustering algorithm. Although many validation metrics exist to estimate the number of clusters, no metric can guarantee that the number of multipath clusters will always be correctly predicted. Most studies continue to use visual recognition to obtain the optimal number of clusters in the environment, which greatly reduces the efficiency of automatic clustering algorithms.

- Most of the clustering algorithms still require many manually inputted pre-defined parameters. For example, in the KPM algorithm, the initial information of clusters (time delay and angle) needs to be defined, and the weight parameters of time delay and angle need to be adjusted iteratively to obtain reasonable output results, and the definition of these parameters is often very subjective. In addition, it is inherently difficult to find reasonable initialization parameters in the actual measurement data. Therefore, it is necessary to establish a multipath clustering algorithm with fewer artificial preset parameters and easier adjustment.

REFERENCES

[1] Molisch A F. *Wireless Communications*. 2nd ed. Chichester: Wiley Publishing, 2011.
[2] He R, Zhong Z, Ai B, et al. Analysis of the relation between Fresnel zone and path loss exponent based on two-ray model. *IEEE Antennas and Wireless Propagation Letters*, 2012, 11(2): 208–211.
[3] Rappaport T S. *Wireless Communications: Principles and Practice*. 2nd ed. Upper Saddle River, NJ: Prentice Hall PTR, 2001.
[4] Keller J B. Geometrical theory of diffraction. *Journal of the Optical Society of America A*, 1962, 52(2): 116–130.
[5] Kouyoumjian R G, Pathak P H. A uniform geometrical theory of diffraction for an edge in a perfectly conducting surface. *Proceedings of the IEEE*, 1974, 62(11): 1448–1461.

[6] Slater S E, Et Frank J C. *Introduction to Theoretical Physics*. New York: McGraw-Hill Book Co., 1933, 315–323.

[7] Bullington K. Radio propagation at frequencies above 30 megacycles. *Proceedings of the IRE*, 1947, 35(10): 1122–1136.

[8] Lee W C Y. Mobile communications engineering. New York: McGraw Hill Publications, 1985.

[9] Sweeney D G, Bostian C W. The dynamics of rain-induced fades. *IEEE Transactions on Antennas and Propagation*, 1992, 40(3): 275–278.

[10] Tatarskii V I. Theory of single scattering by random distributed scatterers. *IEEE Transactions on Antennas and Propagation*, 2003, 51(10): 2806–2813.

[11] Salous S. *Radio Propagation Measurement and Channel Modelling*. 1st ed. Chichester: Wiley Publishing, 2013.

[12] Ament W S. Toward a theory of reflection by a rough surface. *Proceedings of the IRE*, 1953, 41(1): 142–146.

[13] Boithias L. *Radio Wave Propagation*. New York: McGraw-Hill Inc., 1987.

[14] Yang D. 杨大成. 移动传播环境. 北京: 机械工业出版社, 2003.

[15] Goldsmith A. *Wireless Communications*. Cambridge: Cambridge Univ. Press, 2004.

[16] Rowe G B, Williamson A G, Egan B. Mobile radio propagation in Auckland at 76 MHz. *Electronics Letters*, 1983, 19(25): 1064–1065.

[17] Cox D C, Murray R, Norris A. 800 MHz attenuation measured in and around suburban houses. *AT&T Bell Laboratories Technical Journal*, 1984, 63(6): 921–954.

[18] Rappaport T S. Characterization of UHF multipath radio channels in factory buildings. *IEEE Transactions on Antennas and Propagation*, 1989, 37(8): 1058–1069.

[19] Abbas S A, Sheikh A U. On understanding the nature of slow fading in LOS microcellular channels. *The 47th IEEE Vehicular Technology Conference, May 4–7, 1997, Phoenix, AZ, USA*. Piscataway, NJ: IEEE Press, 1997: 662–666.

[20] Abdi A, Kaveh M. On the utility of Gamma PDF in modeling shadowing (slow fading). *The 49th IEEE Vehicular Technology Conference, May 16–20, 1999, Houston, TX, USA*. Piscataway, NJ: IEEE Press, 1999: 2308–2312.

[21] Erceg V, Greenstein L J, Tjandra S Y, et al. An empirically based path loss model for wireless channels in suburban environments. *IEEE Journal on Selected Areas in Communications*, 1999, 17(7): 1205–1211.

[22] Gudmundson M. Correlation model for shadowing in mobile radio systems. *Electronics Letters*, 1991, 27(23): 2145–2146.

[23] Kyösti P, Meinilä J, Hentila L, et al. *Winner II Channel Models*. Chichester: A John Wiley and Sons, Ltd., 2007.

[24] ITU- R M1225. *Guidelines for Evaluation of Radio Transmission Technologies for IMT-2000*. Geneva: ITU, 1997.

[25] Giancristofaro D. Correlation model for shadowing in mobile radio channels. *Electronics Letters*, 1996, 32(11): 958–959.

[26] Algans A, Pedersen K I, Mogensen P E. Experimental analysis of the joint statistical properties of azimuth spread, delay spread, and shadowing. *IEEE Journal on Selected Areas in Communications*, 2002, 20(3): 523–531.

[27] He R, Renaudin O, Kolmonen V M, et al. Characterization of quasi-stationarity regions for vehicle-to-vehicle radio channels. *IEEE Transactions on Antennas and Propagation*, 2015, 63(5): 2237–2251.

[28] Costa N, Haykin S. *Multiple-Input Multiple-Output Channel Models*. Hoboken; Piscataway, NJ: John Wiley & Sons, 2010.

[29] Turin G L, Clapp F D, Johnsoton T L, et al. A statistical model of urban multipath propagation. *IEEE Transactions on Vehicular Technology*, 1972, 21(1): 1–9.

[30] Saleh A M, Valenzuela R A. A statistical model for indoor multipath propagation. *IEEE Journal on Selected Areas in Communications*, 1987, 5(2): 128–137.

[31] Hashemi H, McGuire M, Vlasschaert T. Measurements and modeling of temporal variations of the indoor radio propagation channel. *IEEE Transactions on Vehicular Technology*, 1994, 43(3): 733–737.

[32] Marinier P, Delisle G Y, Despins C L. Temporal variations of the indoor wireless millimeter-wave channel. *IEEE Transactions on Antennas and Propagation*, 1998, 46(6): 928–934.

[33] Laurila J, Hugl K, Toeltsch M, et al. *Directional Wideband 3-D Measurements of Mobile Radio Channel in Urban Environment*. COST 259 Temporary Document TD (99), 1999, 92.

[34] Chong C, Tan C M, Laurenson D, et al. A new statistical wideband spatio-temporal channel model for 5-GHz band WLAN systems. *IEEE Journal on Selected Areas in Communications*, 2003, 21(2): 139–150.

[35] Yu K, Li Q, Cheung D, et al. On the tap and cluster angular spreads of indoor WLAN channels. *The 59th IEEE Vehicular Technology Conference, May 17–19, 2004, Milano, Italy*. Piscataway, NJ: IEEE Press, 2004: 17–19.

[36] Czink N, Yin X, Ozcelik H, et al. Cluster characteristics in a MIMO indoor propagation environment. *IEEE Transactions on Wireless Communications*, 2007, 6(4): 1465–1475.

[37] Steinbauer M, Molisch A, Bonek E. The double-directional radio channel. *IEEE Antennas & Propagation Magazine*, 2001, 43(4): 51–63.

[38] Fleury B, Jourdan P, Stucki A. High-resolution channel parameter estimation for MIMO applications using the SAGE algorithm. *2002 International Zurich Seminar on Broadband Communications, February 19–21, Zurich, Switzerland*. Piscataway, NJ: IEEE Press, 2002.

[39] Fleury B H, Tschudin M, Heddergott R, et al. Channel parameter estimation in mobile radio environments using the SAGE algorithm. *IEEE Journal on Selected Areas in Communications*, 1999, 17(3): 434–450.

[40] IST-2003-507581. WINNER D5.4. Final report on link level and system level channel models. Ver 1.4. October 2005.

[41] Kyösti P, Meinilä J, Hentilä L, et al. WINNER II channel models. IST-WINNER D1.1.2. 2007.

[42] He R, Chen W, Ai B, et al. A sparsity-based clustering framework for radio channel impulse responses. *The 83rd IEEE VTC-Spring, May 15–18, 2016, Nanjing, China*. Piscataway, NJ: IEEE Press, 2016: 1–5.

[43] He R, Li Q, Ai B, et al. An automatic clustering algorithm for multipath components based on Kernel-power-density. *IEEE WCNC, March 19–22, 2017, San Francisco, CA, USA*. Piscataway, NJ: IEEE Press, 2017: 1–6.

[44] Czink N. *The Random-Cluster Model – A Stochastic MIMO Channel Model for Broadband Wireless Communication Systems of the 3rd Generation and Beyond*. Vienna: Technische Universitat Wien, 2007.

[45] Czink N, Cerap, Saloj, et al. A framework for automatic clustering of parametric MIMO channel data including path powers. *The 64th IEEE Vehicular Technology Conference, September 25–28, Montreal, Que., Canada*. Piscataway, NJ: IEEE Press, 2006: 1–5.

[46] Schneider C, Bauer M, Narandzic M, et al. Clustering of MIMO channel parameters-performance comparison. *The 69th IEEE Vehicular Technology Conference, April 26–29, Barcelona, Spain*. Piscataway, NJ: IEEE Press, 2009: 1–5.

[47] Wang Q, Ai B, He R, et al. A framework of automatic clustering and tracking for time-variant multipath components. *IEEE Communications Letters*, 2017, 21(4): 953–956.

CHAPTER 2

Wireless Channel Measurement Technology

2.1 SIGNIFICANCE OF CHANNEL SOUNDING

Performance of a wireless communication system is mainly restricted by a wireless channel. In a mobile communication system, the propagation path between the transmitter and receiver is complex, ranging from simple LOS propagation to various complex terrains, such as large buildings and mountains. The received signal has strong randomness in different physical environments, which makes it difficult to directly obtain signal strength of each location in theory and to analyze channel characteristics. In addition, mobility of the transmitter and receiver will also affect fading characteristics. Modeling of wireless channels has always been challenging in design of wireless communication systems. Therefore, when investigating radio wave propagation characteristics of mobile communication, wireless channel measurement is one of the essential and important approaches. When making a link budget, it is first necessary to obtain various empirical formulas and models for radio wave propagation calculations through statistical analysis of a large number of measured data, combining with the characteristics of terrain and ground objects, and then to make more accurate predictions. However, due to the significant differences in the terrain and buildings in the actual environment, it is necessary to use channel measurements to verify the degree of conformity between the actual situation and the radio wave propagation prediction model.

Wireless channel measurement can also provide raw data for wireless channel modeling, which is convenient for researchers to use statistical methods to complete the description of the wireless signal change process of different communication systems under different propagation environments. On the one hand, the data obtained by channel measurement can be used to obtain channel parameters, so as to realize statistical channel modeling; on the other hand, it can also be used to compare with the simulation data obtained by using electromagnetic field theory, so as to carry out deterministic channel modeling.

DOI: 10.1201/9781032669793-2

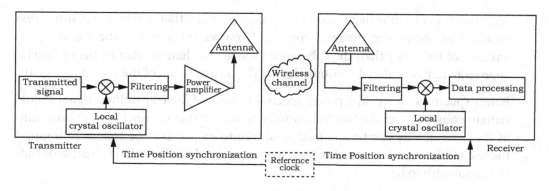

FIGURE 2.1   General channel measurement system structure.

It can be seen that to conduct the analysis and research of the radio wave propagation characteristics of mobile communications, it is very necessary to study high-accuracy and efficient channel measurement methods in advance.

Wireless channel measurement is also called channel sounding, that is, the transmitter sends a known training signal, and the receiver stores the received signal and estimates information of wireless channel through the known transmitted signal. According to the processing of received signal, channel impulse response or the corresponding channel transfer function is obtained.

As communication systems become more and more complex, more accurate and practical channel models and channel sounding techniques are required. In the 1960s, with the introduction of broadband wireless communication systems, channel sounders were able to record channel impulse response. In the 1990s, directional propagation channel characteristics gradually attracted more attention. Channel detection gradually began to use multiple transmitting and receiving antennas and more complex antenna arrays to perform bidirectional channel detection and serve the establishment of the directional channel model.

The structure of a general wireless channel measurement system is shown in Figure 2.1 [1], which is mainly composed of a signal source, a local crystal oscillator, a radio frequency (RF) link, a power amplifier, an intermediate frequency filter, a signal processing unit, and an antenna. The transmitter sends a radio frequency signal at a specific carrier frequency and acts on the wireless channel, and the receiver can identify and detect the transmitted signal after channel attenuation and then store it.

The performance of the detector mainly depends on the choice of the detection signal. To achieve an effective channel measurement, the design of the transmitted signal often depends on the following factors [1].

- Signal bandwidth: The transmitted signal bandwidth is inversely proportional to its minimum time variation, thus determining the maximum delay resolution of the detection channel.

- Signal duration: The effective duration of the signal must be adapted to the corresponding channel characteristics. The longer the signal duration, the more energy is sent under a certain transmission power condition, so that a higher signal-to-noise ratio can be obtained at the receiving end. However, the length of the sounding signal cannot be longer than the coherence time of the channel, that is, the channel is approximately considered constant during the transmission of the sounding signal.

- Power spectral density: The power spectral density of the transmitted signal should remain constant over the bandwidth to be tested, so that channel estimation results of the same quality can be obtained at all frequency points to be tested. To improve the test efficiency, the transmitted signal energy should be as little as possible outside the bandwidth to be tested.

- Peak factor: Peak factor is defined as the ratio of the peak amplitude of the transmitted signal to its root mean square amplitude. A lower peak factor will help the transmit power amplifier to function efficiently.

2.2 TIME DOMAIN CHANNEL SOUNDING

Channel sounding based on time domain refers to a direct measurement of time domain impulse response of the channel. Assuming that the channel is slowly time-varying, the measured impulse response is equal to the convolution of the actual channel impulse response and the impulse response of the channel measurement system, which can be expressed as:

$$h_{\mathrm{meas}}(t,\tau) = p(\tau) \times h(t,\tau) \qquad (2.1)$$

where $p(\tau)$ represents the impulse response of the channel measurement system, and $h(t,\tau)$ represents the actual channel impulse response. Equation (2.1) is obtained under the assumption that both the channel and the transceiver are linear. The impulse response of the wireless channel measurement system should be as close to the ideal δ function as possible to minimize the influence of the measurement system on the channel measurement results.

2.2.1 Pulse-Based Channel Sounding Technique

Channel sounding with RF pulses is a simple way of time-domain channel measurement. A direct RF pulse measurement means that the transmitter simulates an ideal impulse function with a very narrow pulse that is repeatedly sent, as shown in Figure 2.2. This technique can quickly measure the power delay distribution of the channel [2–4]. The receiver adopts a band-pass filter to receive the narrow pulse signal, so as to band-limit and filter it. After being detected by the envelope detector, it is amplified, stored, and processed, and finally, the amplitude information of the impulse response is obtained.

The advantage of this method is that it is simple in principle, easy to implement, and can quickly obtain the power delay spectrum of channel multipath. The disadvantage is that both the transmitting and receiving signal ends are susceptible to interference from other signals of the same frequency. When the power of the transmitted signal is limited, the strength of the obtained multipath signal is low, so the measurement performance

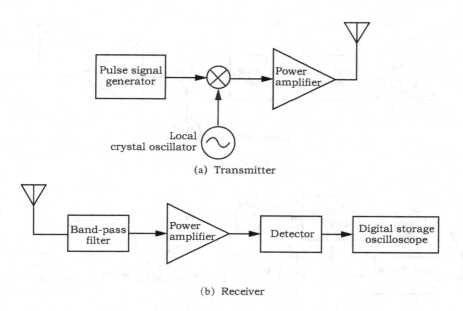

(a) Transmitter

(b) Receiver

FIGURE 2.2 Direct RF channel impulse response measurement system.

mainly depends on the performance of the narrow pulse and the receiving sensitivity of the receiver. In addition, this scheme cannot extract the phase characteristic information of the channel, so it cannot be combined with the MIMO array to realize the extraction of multipath angle domain information in the test data.

2.2.2 Correlation-Based Channel Sounding Technique

Channel detection technology based on sliding correlation has high measurement accuracy, and the spread spectrum sliding correlation method can be applied to time-varying environments, so it has been widely used. Most channel detectors today use the principle of correlation channel measurement [5].

In a spread spectrum correlation detector, it is first necessary to determine the pseudo noise sequence used by the system. During the measurement process, the transmitting end sends this specific sequence as a measurement signal cyclically, and the receiving end saves this sequence as a local sequence, and continues to perform correlation calculations with the received signal. Due to the good autocorrelation of the pseudo-random sequence, the method can effectively detect the multipath signal in the channel. Only when the local pseudo-random sequence is completely aligned with the originator, the correlation operation can obtain the maximum output peak. The local chips at the receiving end continuously slide relative to each other, so as to obtain several maximum correlation outputs, and finally determine the time delay values of all multipath signals. The basic principle of this method is shown in Figure 2.3.

Correlation-based channel detection technology has many advantages: on the one hand, the receiver of the system can use a wide-band mixer plus a narrow-band receiver to detect wide-band transmission signals; on the other hand, in the spread spectrum sliding

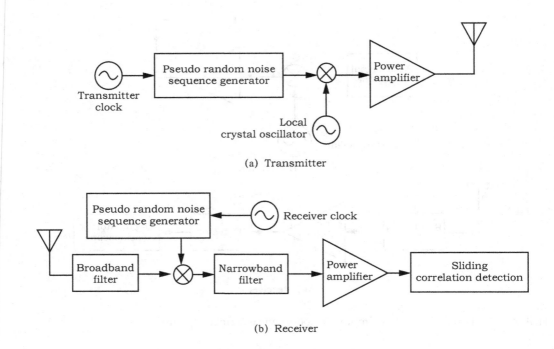

(a) Transmitter

(b) Receiver

FIGURE 2.3   Spread spectrum correlation channel measurement system.

correlation channel system, one of its spread spectrum modulation characteristics is that it can filter out band-pass interference and improve the coverage of a given transmit power. Correlation-based channel detection technology also utilizes the good autocorrelation of pseudo-random sequences to reduce the average noise power and average interference power in the receiving bandwidth, and has good anti-interference characteristics.

This sounding technique increases the dynamic range of the system compared to pulse-based channel sounding techniques. But its disadvantage is that the measurement and processing results are not real-time, and the measurement time of the power delay distribution depends on the system parameters and measurement purpose, and sometimes it is too long. Another disadvantage is that using a non-coherent detector will not detect the respective phases of the multipath components, and even with a coherent detector, the sweep time of the spread-spectrum signal will cause a large delay, so that the phases of the multipath components with different delays are detected at different times, but the channel cannot be guaranteed to be stable.

2.3 FREQUENCY DOMAIN CHANNEL SOUNDING

Since there is a corresponding relationship between the frequency domain and time domain, the transfer function of the channel can also be measured in frequency domain, so as to realize the purpose of wireless channel sounding. The main design criteria of the transmitted signal waveform in the frequency domain measurement are as follows: in the bandwidth to be measured, the detection signal should have a flat power spectrum and be able to support the analysis of the measurement results directly in the frequency domain.

Common frequency domain channel detectors include vector network testers, short pulse channel detectors, etc. Figure 2.4 shows a frequency domain detector used to detect channel transfer function, including a vector network tester with frequency sweep function, transmitting antenna, and receiving antenna.

The vector network analyzer (VNA) is used to measure the parameters of the wireless channel, and the transfer function of the channel is the value of the parameter at a certain frequency. For each specific frequency, the VNA sends a known signal from port 1 and receives a signal level from port 2. The channel transfer function H (that is, the calculated S_{21} parameters) within the test frequency range can be obtained by making the excitation signal sweep or step change over the entire frequency range to be tested. The response here is the expression of the channel impulse response in the frequency domain, which is transformed into a time domain expression by the inverse discrete Fourier transform (IDFT), so as to obtain the time domain impulse response of the channel.

Usually, the measurement result obtained by using a VNA is relatively accurate, and the channel can be measured directly. This technique can provide the amplitude and phase information of the channel in time domain. However, in actual measurements, such a sounding system requires precise time positioning and strict synchronization at the transceiver end, which limits the test distance to a certain extent, and the transmitter and receiver are in the same physical device. These all determine that the frequency domain measurement based on the VNA is mainly suitable for indoor and outdoor short-distance channel measurements.

Also, another limitation of this type of measurement is the non-real-time nature of the test. Due to the long measurement process, the repeated measurement frequency generally cannot exceed several Hz. For time-varying channels, the frequency response changes quickly, and it is difficult to ensure that the channel does not change significantly during a channel measurement process. Therefore, this method is mainly used for channel measurement in a static environment.

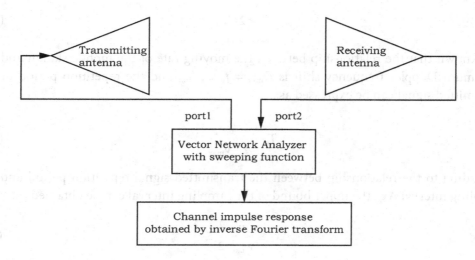

FIGURE 2.4   Frequency domain channel measurement system.

2.4 STATIC AND DYNAMIC CHANNEL SOUNDING

According to the characteristics of the channel, measurements can be divided into static channel measurement and dynamic channel measurement. Static testing is not covered in detail in this book. Common dynamic channel measurement scenarios include vehicle communication scenarios, rail transit communication scenarios, UAV communication scenarios, etc. Dynamic wireless channels usually have significant time-varying characteristics, which need to be described by non-stationary fading processes. Since the channel is time-varying rapidly, the channel detector needs to adapt to the channel measurement requirements in terms of sampling interval configuration [6] and fast storage.

The time-varying nature of wireless channel has a certain impact on whether the channel can be uniquely identified (measured). For a time-invariant band-limited channel, one of the important principles for the receiver to accurately restore the transmitted signal is to satisfy the Nyquist theorem in the delay domain [7]. That is, assuming that the bandwidth of the band-pass signal is B, and the highest cut-off frequency is f_h, then when the sampling frequency f_s satisfies $f_s = 2f_H / m$, the transmitted signal can be reconstructed from the sampling sequence without distortion. When the signal center carrier frequency f_c is much larger than the signal bandwidth B, the sampling rate is approximately equal to $2B$.

In a time-varying system, the channel response of any impulsive excitation signal can be regarded as a "snapshot" of the channel. Obviously, in order to track the change of the wireless channel, the repetition period T_s of the probe signal must be less than the time when the channel changes significantly. Just as a minimum sampling frequency is required to distinguish time-invariant band-limited signals, a minimum time-sampling frequency is also required to identify time-varying processes with band-limited Doppler spectra. Therefore, the temporal sampling frequency is at least twice the maximum Doppler shift D_{max} [1], which can be represented as:

$$f_s \geq 2D_{max} \tag{2.2}$$

It is known that the relationship between the moving rate of the mobile station and the maximum Doppler frequency shift is $D_{max} = f_c v_{max}/c_0$, and the repetition period of the transmitted signal can be expressed as:

$$T_s \leq \frac{c_0}{2f_c v_{max}} \tag{2.3}$$

According to the relationship between the transmitted signal repetition period and the sampling interval Δx_s, the upper bound of the sampling interval can be obtained as:

$$\Delta x_s \leq \frac{v}{v_{max}} \frac{\lambda}{2} \leq \frac{\lambda}{2} \tag{2.4}$$

Generally, for a narrowband channel measurement, in order to fully reflect the impact of small-scale fading on channel characteristics, a sampling interval of 1/4 wavelength is often used.

In a high-speed mobile wireless system, the high-speed movement of the vehicle receiver will cause rapid changes in the radio wave propagation environment where the communication terminal is located. Usually, delay spread and coherent bandwidth are used to describe the time dispersion characteristics of the local channel, and Doppler spread and coherent time are used to describe the time-varying characteristics of the small-scale channel. This time-varying characteristic is caused by the relative motion between the mobile station and the base station or by the motion of other objects in the channel.

Doppler spread B_D is a measure of the extent of spectral broadening caused by the time-varying rate of a mobile radio channel. Doppler spread is the frequency range over which the received Doppler spectrum has non-zero values. When a pure sinusoidal signal with frequency f_c is transmitted, the spectrum of the received signal (i.e., the Doppler spectrum) has components in the range of $f_c - f_d$ and $f_c + f_d$, where f_d is the Doppler shift. The Doppler frequency shift is a function of the relative velocity of the mobile station, the angle between the direction of motion of the mobile station, and the arrival direction of the scattered wave. If the baseband channel bandwidth is much larger than B_D, the effect of Doppler spread can be ignored at the receiver.

Coherence time T_c is the expression of Doppler spread in the time domain, which is used to describe the time-varying characteristics of channel frequency dispersion in the time domain. The coherence time is the statistical average of the time interval during which the channel impulse response remains unchanged, that is, within a specific time range, the signals arriving at different times have a strong amplitude correlation. Coherence time is inversely proportional to Doppler shift. If the coherence time is defined as the length of the time period for which the time correlation function is greater than 0.5, the coherence time is approximately expressed as:

$$T_c \approx \frac{9}{16\pi D_{\max}} \tag{2.5}$$

Through the coherence time, the coherence distance can also be obtained, that is, the statistical characteristics of channel small-scale fading can assume a constant maximum distance interval.

For severely time-varying channels, on the one hand, to ensure that the impulse responses of different excitation pulse signals do not overlap, the transmitted signal repetition period T_s must be greater than the maximum additional delay $\tau_{\max}$ of the channel; on the other hand, the repetition period must satisfy $T_s \leq \frac{1}{2D_{\max}}$. Therefore, only when the channel parameters satisfy equation (2.6) can the accurate measurement of the channel be realized.

$$2\tau_{\max} D_{\max} \leq 1 \tag{2.6}$$

2.5 DYNAMIC CHANNEL SOUNDING CAMPAIGN

Dynamic channel sounding generally includes single-ended mobile dynamic channel sounding and double-ended mobile dynamic channel sounding. Single-ended mobile dynamic channel sounding activities mainly include urban mobile cellular channel sounding, vehicle-to-infrastructure (V2I) roadside communication unit sounding, high-speed railway channel sounding, and air-to-ground channel sounding; double-ended mobile dynamic channel sounding activities mainly include vehicle-to-vehicle (V2V) communication channel sounding, device-to-device (D2D) communication channel sounding, etc.

In recent years, some research institutions at home and abroad have carried out a large number of dynamic channel sounding in the corresponding dynamic communication scenarios by using industry-recognized wireless channel sounders or wireless channel measurement equipment.

2.5.1 V2V Channel Sounding at Aalto University in Finland

The multi-antenna dynamic channel test platform independently designed by Aalto University in Finland is based on the principle of array switching, and realizes sequential capture of the MIMO multi-antenna and multiple link signals through fast array switching on both sides of the transmitting station and receiving station [8]. The Aalto channel tester works in the 5.3 GHz frequency band, the maximum test bandwidth of the channel tester can support 120 MHz, the receiver sampling rate can reach 120 MHz, and the transmit power is 36 dBm. In channel measurement, a pseudo-random noise sequence with a length of 4.25 μs is used as the transmitted signal. In the test, it takes 1.632 ms to complete the acquisition of all MIMO antenna link signals. The channel tester can be set to time trigger mode with a sampling interval of 15 ms. For a more detailed description of the Aalto University channel tester, please refer to refs. [9,10], and its main specifications are shown in Table 2.1.

Figure 2.5 shows the antenna array of the dynamic MIMO test system. The receiving station uses a dual-polarized hemispherical antenna array. The hemispherical array contains 15 dual-polarization (horizontal and vertical polarization) antenna elements, that is, there are 30 signal receiving links. The hemispherical antenna array can estimate the three-dimensional spatial angle characteristics of the wireless channel through the phase

TABLE 2.1   Aalto University Channel Sounder Parameters

Parameters	Value
Carrier frequency	5.3 GHz
Bandwidth	10~200 MHz
Transmitted signal type	Pseudo noise sequence
Transmitted signal length	1.6~25.6 μs
The maximum sampling rate of the transmitted signal	320 MHz
Receiver maximum sampling rate	400 MHz
Receive antenna switching interval	3.2/6.4 μs
MIMO snapshot interval	39.32/72.09 ms
MIMO channel single measurement time	3.277/6.554 ms

FIGURE 2.5   Photo of a hemispherical antenna array in a dynamic MIMO test system [9].

difference of the received signal in the horizontal and vertical planes. The transmitting station adopts a vertically polarized uniform linear antenna array (composed of four vertically polarized antennas). For channel measurements, the antenna arrays of the transmitter and receiver stations are mounted on a wooden platform on top of the car.

The channel tester of Aalto University in Finland has completed the channel measurement for urban areas [11–14], suburbs [15], underground parking lots [16], tunnels [17], and intersections in V2V environments [18,19] in traditional cellular networks. In the above tests, dynamic V2V channel measurements were carried out in the city of Tapiola, Finland, and its surrounding suburbs. The dynamic measurement scenarios are divided into suburbs, urban areas, underground car park tunnels (located underground in urban areas), and intersections (located in suburban areas). There are some discrete low houses, parking lots, and a little vegetation with a height of about 5 m in the measured suburban area. The measured road section is a two-way lane, and there are wider sidewalks and sporadic traffic signs in the two sections of the lane. There are a large number of 3–4 storey buildings in the measured urban area, and there are many traffic signs and vegetation around the two-way lanes. The underground car park tunnel is located underground in the city center section of Tapiola, and there are a large number of scatterers in the tunnel. The illustrations of the above scenarios can be found in ref. [10].

2.5.2 V2V Channel Sounding at Lund University in Sweden

Lund University in Sweden has a complete RUSK MIMO broadband channel detector. Its multi-antenna switching principle is similar to that of the Aalto University channel detector. Through fast array switching on both sides of the transmitting station and receiving station, the sequential acquisition of MIMO multi-antenna and multiple link signals is realized [20], as shown in Figure 2.6. The transmitter can generate multi-carrier periodic signals of arbitrary waveforms, and its main specifications are shown in Table 2.2.

Using this channel tester, researchers at Lund University have conducted V2V channel measurement activities in different radio wave propagation scenarios many times. Based on the European Telecommunications Standards Institute (ETSI), the Lund University research team mainly conducted channel detection in the following specific V2V application scenarios:

- Different types of intersections [21–24];

- Merging lane [25–27];

- Expressways with LOS propagation conditions [28–31];

- Traffic congestion (approaching or overtaking in traffic jams) [32];

- Tunnels [33–36].

FIGURE 2.6   Photo of V2V channel measurement system.

TABLE 2.2 Lund University Channel Sounder Parameters

Parameters	Value
Carrier frequency	0.3/2/5 GHz
Bandwidth	10~240 MHz
Transmitted signal type	Multi-carrier signal
Transmitted signal length	1.6~25.6 μs
Transmitted signal sampling rate	320 MHz
Receiver sampling rate	640 MHz
Receive antenna switching interval	3.2/6.4 μs
MIMO snapshot interval	39.32/72.09 ms
MIMO channel single measurement time	3.277/6.554 ms

(a) motorway

(b) suburban

(a) crossroad

(b) tunnel

FIGURE 2.7 Example of V2V dynamic measurement scenarios.

Figure 2.7 shows the above-mentioned V2V dynamic channel test environment.

2.5.3 Channel Sounding at University of South Carolina in US

The channel tester developed by the research team of the University of South Carolina in the United States is a commercial channel tester designed by the spread spectrum method. The channel tester uses a direct spread spectrum sequence $c(t)$ to modulate the sin function carrier to achieve the purpose of broadening the spectrum. This spreading sequence $c(t)$ is a pseudo-random noise sequence in the channel tester. Any two of these multipath components can be resolved by a minimum delay so that they can be unambiguously distinguished by the receiver. This minimum delay is called the delay resolution of the wireless channel tester. Each chip $c(t)$ is similar to a pulse signal, so the delay resolution capability of the multipath component of the wireless channel tester is determined by the duration T_c of the chip. At the receiving end, the channel tester adopts a method called "step correlation." The processing process is that the receiver performs correlation calculations on the received signal at each time delay τ within a period of time, and outputs the results; and then moves the delay time T_c, performs related calculations, and outputs the results. This way, the channel tester probes over the entire unambiguous delay interval.

In addition, the carrier clock synchronization problem of the transceiver is critical to the transmitter and receiver step correlation processing. Therefore, the transceivers must be connected by coaxial cable to train them synchronously before the test begins. The receiver impulse response output is essentially a convolution of the channel impulse response and the autocorrelation of the chip signal. Figure 2.8 shows the actual channel tester used in the test work. Table 2.3 lists the main parameters of the channel tester.

The scenarios tested by the channel sounder of the University of South Carolina research team mainly include typical V2V scenarios (urban areas, suburbs, etc.) [37–41] and atypical V2V complex scenarios (slope roads [42], parking lots [43–45], overpasses [46,47], etc.).

2.5.4 High-Speed Railway Channel Sounding

The channel measurement activities carried out in the high-speed railway channel environment are introduced as follows.

FIGURE 2.8 Photo of the channel tester at the University of South Carolina.

TABLE 2.3   University of South Carolina Channel Sounder Parameters

Parameters	Value
Output power	6~33 dBm
Center frequency	5.095~5.20 GHz
Chip rate	50 Mchip/s
99% bandwidth range	52.76 MHz
Receiver sampling frequency	100 MHz
Test rate	2~60 PDP/s
Transmit antenna	Omnidirectional antenna, 8.5 dBi
Receive antenna	Omnidirectional antenna, 17 dBi

In 2006, German MEDAV, Karlsruhe University and Ilmenau University of Technology used the standard channel measurement equipment RUSK channel sounder for the first time to carry out wireless channel measurement on the ICE high-speed railway in Siegburg and Frankfurt, Germany, measured wireless channel propagation characteristics in a high-speed railway scenario [48]. The principle of the channel sounder is the same as that of the Lund University channel sounder, and the parameters are similar, which will not be repeated here.

In 2008, Finland's Elektrobit and the University of Oulu, together with the Taiwan Industrial Technology Research Institute, used Propsound to implement high-speed railway channel measurements in rural and mountainous areas on the Hsinchu-Taichung and Taichung-Chiayi lines of Taiwan's high-speed railway, respectively. The small-scale fading characteristics of the channel were obtained, and the characteristics of the Doppler rapid change of the high-speed railway channel were verified [49]. In 2010, Beijing University of Posts and Telecommunications and China Mobile Research Institute used Propsound to conduct channel measurements on the internal environment of high-speed train carriages at the Tangshan Bus Factory in China. Parameters such as path loss, delay spread, and Ricean distribution K factor of the propagation environment inside the car are extracted [50]. In 2012, the Institute of Broadband Wireless Mobile Communication of Beijing Jiaotong University cooperated with China Mobile Research Institute to realize the channel measurement of the Beijing-Tianjin high-speed railway, and analyzed the large-scale and small-scale fading characteristics of the channel under the unobstructed viaduct scene [51]. In 2012, Beijing Jiaotong University and China Mobile Research Institute continued to carry out channel measurement [52–55] on the Zhengxi high-speed railway that shielded viaducts and U-shaped troughs. Different from the Beijing-Tianjin intercity high-speed rail measurement scene, there are trees blocking the wireless link of the viaduct scene in this measurement.

Here we focus on the Propsound channel detector [56] from Elektrobit. The Propsound channel sounding system consists of a transmitter and a receiver, as shown in Figure 2.9. It is suitable for channel sounding activities in multiple frequency bands, and can support single-antenna and multi-antenna broadband tests. The channel sounder is based on time-delay domain spread spectrum technology, and its multi-antenna channel measurement scheme design is also based on fast-switching antenna arrays. The receiver is connected to the data acquisition module to directly obtain channel parameters such as channel impulse response. See Table 2.4 for detailed instrument parameters.

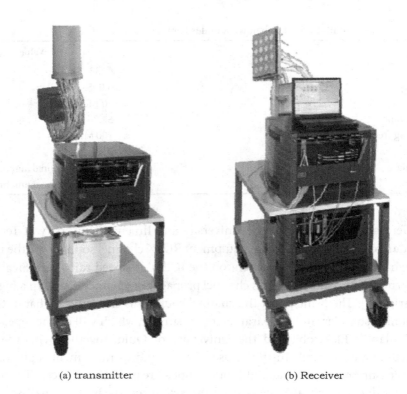

(a) transmitter   (b) Receiver

FIGURE 2.9   Propsound channel tester in kind.

TABLE 2.4   Propsound Channel Detector Parameters

Parameters	Value
Carrier frequency	1.7~2.1 GHz, 2.0~2.7 GHz, 3.2~4.0 GHz, 5.1~5.9 GHz
Bandwidth	1.56~200 MHz
Transmitted signal power	26 dBm
Chip length	255
Test rate	30,000 CIR/s
Chip rate	100 Mchip/s
Chip length	31~4,095
Measurement mode	SISO, SIMO, MIMO
Baseband sampling rate	2 GS/s
Synchronous mode	Rubidium clock

REFERENCES

[1] Molisch A F. *Wireless Communications*. 2nd ed. Hoboken, NJ: Wiley, 2010.
[2] Rappaport T S. Characterization of UHF multipath radio channels in factory buildings. *IEEE Transactions on Antennas and Propagation*, 1989, 37(8): 1058–1069.
[3] Rappaport T S, Seidels Y, Singh R. 900-MHz multipath propagation measurements for US digital cellular radiotelephone. *IEEE Transactions on Vehicular Technology*, 1990, 39(2): 132–139.
[4] He R, Zhong Z, Ai B, et al. Propagation channel measurements and analysis at 2.4 GHz in subway tunnels. *IET Microwaves, Antennas & Propagation*, 2013, 7(11): 934–941.

[5] Rappaport T S. *Wireless Communications: Principles and Practice.* 2nd ed. Upper Saddle River, NJ: Prentice Hall PTR, 2001.

[6] Wang T, Ai B, He R, et al. Local mean power estimation over fading channels. *IEEE VTC-Spring, May 15–18, 2016, Nanjing, China.* Piscataway, NJ: IEEE Press, 2016: 1–5.

[7] Ziemer R E, Tranter W H. *Principles of Communications: Systems, Modulation and Noise.* 7th ed. New York: Wiley, 2013.

[8] Kolmonen V, Kivinen J, Vuokko L, et al. 5.3 GHz MIMO radio channel sounder. *IEEE Transactions on Instrumentation and Measurement,* 2006, 55(4): 1263–1269.

[9] Renaudin O, Kolmonen V, Vainikainen P, et al. Non-stationary narrowband MIMO inter-vehicle channel characterization in the 5-GHz band. *IEEE Transactions on Vehicular Technology,* 2010, 59(4): 2007–2015.

[10] Renaudin O. *Experimental Channel Characterization for Vehicle-to-Vehicle Communication Systems.* Belgium: Universit'e Catholique de Louvain, 2013.

[11] Renaudin O, Kolmonen V, Vainikainen P, et al. Wideband MIMO car-to-car radio channel measurements at 5.3 GHz. *The 68th IEEE Vehicular Technology Conference, September 21–24, 2008, Calgary, BC, Canada.* Piscataway, NJ: IEEE Press, 2008: 1–5.

[12] Renaudin O, Kolmonen V, Vainikainen P, et al. Car-to-car channel models based on wideband MIMO measurements at 5.3 GHz. *The 3rd European Conference on Antennas and Propagation, March 23–27, 2009, Berlin, Germany.* Piscataway, NJ: IEEE Press, 2009: 635–639.

[13] He R, Renaudin O, Kolmonen V M, et al. Characterization of quasi-stationarity regions for vehicle-to-vehicle radio channels. *IEEE Transactions on Antennas and Propagation,* 2015, 63(5): 2237–2251.

[14] He R, Renaudin O, Kolmonen V M, et al. A dynamic wideband directional channel model for vehicle-to-vehicle communications. *IEEE Transactions on Industrial Electronics,* 2015, 62(12): 7870–7882.

[15] He R, Renaudin O, Kolmonen V M, et al. Non-stationarity characterization for vehicle-to-vehicle channels using correlation matrix distance and shadow fading correlation. *Progress in Electromagnetics Research Symposium, August 25–28, 2014, Guangzhou, China.* Cambridge, MA: Electromagnetics Academy, 2014: 2144–2148.

[16] He R, Renaudin O, Kolmonen V M, et al. Statistical characterization of dynamic multi-path components for vehicle-to-vehicle radio channels. *IEEE 81st Vehicular Technology Conference (VTC Spring), May 11–14, 2015, Glasgow, United Kingdom.* Piscataway, NJ: IEEE Press, 2015: 1–6.

[17] Aikio P, Gruber R, Vainikainen P. Wideband radio channel measurements for train tunnels. *48th IEEE Vehicular Technology Conference (VTC'98), May 21, 1998, Ottawa, ON, Canada.* Piscataway, NJ: IEEE Press, 1998: 460–464.

[18] He R, Renaudin O, Kolmonen V M, et al. Vehicle-to-vehicle radio channel characterization in cross-road scenarios. *IEEE Transactions on Vehicular Technology,* 2016, 65(8): 5850–5861.

[19] He R, Renaudin O, Kolmonen V M, et al. Angular dispersion characterization of vehicle-to-vehicle channel in cross-road scenarios. *European Conference on Antennas and Propagation (EuCAP), April 13–17, 2015, Lisbon, Portugal.* Piscataway, NJ: IEEE Press, 2015: 1–4.

[20] Thomä R, Hampicke D, Richter A, et al. Identification of time-variant directional mobile radio channels. *IEEE Transactions on Instrumentation and Measurement,* 2000, 49(2): 357–364.

[21] Paier A, Karedal J, Czink N, et al. Car-to-car radio channel measurements at 5 GHz: Pathloss, power-delay profile, and delay-Doppler spectrum. *2007 4th International Symposium on Wireless Communication Systems, October 17–19, 2007, Trondheim, Norway.* Piscataway, NJ: IEEE Press, 2007: 224–228.

[22] Paier A, Karedal J, Czink N, et al. First results from car-to-car and car-to-infrastructure radio channel measurements at 5.2GHz. *2007 IEEE 18th International Symposium on Personal, Indoor and Mobile Radio Communications, September 3–7, 2007, Athens, Greece.* Piscataway, NJ: IEEE Press, 2007: 1–5.

[23] Karedal J, Tufvesson F, Czink N, et al. A geometry-based stochastic MIMO model for vehicle-to-vehicle communications. *IEEE Transactions on Wireless Communications*, 2009, 8(7): 3646—3657.

[24] Karedal J, Tufvesson F, Abbas T, et al. Radio channel measurements at street intersections for vehicle-to-vehicle safety applications. *2010 IEEE 71st Vehicular Technology Conference (VTC 2010-Spring), May 16–19, 2010, Taipei, Taiwan, China*. Piscataway, NJ: IEEE Press, 2010: 1—5.

[25] Abbas T, Bernado L, Thiel A, et al. Radio channel properties for vehicular communication: Merging lanes versus urban intersections. *IEEE Vehicular Technology Magazine*, 2013, 8(4): 27—34.

[26] Abbas T, Bernado L, Thiel A, et al. Measurements based channel characterization for vehicle-to-vehicle communications at merging lanes on highway. *2013 IEEE 5th International Symposium on Wireless Vehicular Communications (WiVeC), June 2–3, 2013, Dresden, Germany*. Piscataway, NJ: IEEE Press, 2013: 1—5.

[27] Abbas T, Karedal J, Tufvesson F. Measurement-based analysis: The effect of complementary antennas and diversity on vehicle-to-vehicle communication. *IEEE Antennas and Wireless Propagation Letters*, 2013, 12(1): 309—312.

[28] Karedal J, Czink N, Paier A, et al. Path loss modeling for vehicle-to-vehicle communications. *IEEE Transactions on Vehicular Technology*, 2011, 60(1): 323—328.

[29] Abbas T, Tufvesson F. Line-of-sight obstruction analysis for vehicle-to-vehicle network simulations in a two lane highway scenario. *International Journal of Antennas and Propagation* 2013, 2013: 459323.

[30] Bernadó L, Zemen T, Tufvesson F, et al. Delay and Doppler spreads of nonstationary vehicular channels for safety-relevant scenarios. *IEEE Transactions on Vehicular Technology*, 2014, 63(1): 82—93.

[31] Bernadó L, Zemen T, Tufvesson F, et al. Time- and frequency-varying *K*-factor of nonstationary vehicular channels for safety-relevant scenarios. *IEEE Transactions on Intelligent Transportation Systems*, 2015, 16(2): 1007—1017.

[32] Paier A, Bernadó L, Kåredal J, Klemp O, Kwoczek A. Overview of vehicle-to-vehicle radio channel measurements for collision avoidance applications. *2010 IEEE 71st Vehicular Technology Conference (VTC 2010-Spring), May 16–19, 2010, Taipei, Taiwan, China*. Piscataway, NJ: IEEE Press, 2010: 1—5.

[33] Mecklenbräuker C, Molisch A, Karedal J, et al. Vehicular channel characterization and its implications for wireless system design and performance. *Proceedings of the IEEE*, 2011, 99(7): 1189—1212.

[34] Abbas T, Karedal J, Tufvesson F, Paier A, Bernado L, Molisch A F. Directional analysis of vehicle-to-vehicle propagation channels. *2011 IEEE 73rd Vehicular Technology Conference (VTC Spring), May 15–18, 2011, Yokohama, Japan*. Piscataway, NJ: IEEE Press, 2011: 1—5.

[35] Abbas T, Nuckelt J, Kürner T, et al. Simulation and measurement-based vehicle-to-vehicle channel characterization: Accuracy and constraint analysis. *IEEE Transactions on Antennas and Propagation*, 2015, 63(7): 3208—3218.

[36] Abbas T. *Measurement Based Channel Characterization and Modeling for Vehicle-to-Vehicle Communications*. Sweden: Lund University, 2014.

[37] Matolak D, Sen I, Xiong W, et al. 5 GHz wireless channel characterization for vehicle to vehicle communications. *IEEE Military Communications Conference, October 17–20, 2005, Atlantic, NJ, USA*. Piscataway, NJ: IEEE Press, 2005: 1—7.

[38] Matolak D W, Sen I, Xiong W. Channel modeling for V2V communications. *2006 Third Annual International Conference on Mobile and Ubiquitous Systems: Networking & Services, July 17–21, 2006, San Jose, CA, USA*. Piscataway, NJ: IEEE Press, 2006: 1—7.

[39] Matolak D W. Channel modeling for vehicle-to-vehicle communications. *IEEE Communications Magazine*, 2008, 46(5): 76—83.

[40] Sen I, Matolak D W. Vehicle-vehicle channel models for the 5-GHz band. *IEEE Transactions on Intelligent Transportation Systems*, 2008, 9(2): 235–245.

[41] Wu Q, Matolak D W, Sen I. 5 GHz band vehicle-to-vehicle channels: Models for multiple values of channel bandwidth. *IEEE Transactions on Vehicular Technology*, 2010, 59(5): 2620–2625.

[42] Liu P, Matolak D W, Ai B, et al. Path loss modeling for vehicle-to-vehicle communication on a slope. *IEEE Transactions on Vehicular Technology*, 2014, 63(6): 2954–2958.

[43] Sun R, Matolak D W, Liu P Y. Parking garage channel characteristics at 5 GHz for V2V applications. *IEEE 78th Vehicular Technology Conference (VTC Fall), September 2–5, 2013, Las Vegas, NV, USA*. Piscataway, NJ: IEEE Press, 2013: 1–5.

[44] Matolak D W, Sun R, Liu P Y. V2V channel characteristics and models for 5 GHz parking garage channels. *2015 9th European Conference on Antennas and Propagation (EuCAP), April 13–17, 2015, Lisbon, Portugal*. Piscataway, NJ: IEEE Press, 2015: 1–4.

[45] Sun R, Matolak D W, Liu P Y. 5GHz V2V channel characteristics for parking. *IEEE Transactions on Vehicular Technology*, 2017, 66(5): 3538–3547.

[46] Liu P, Ai B, Matolak D W, Sun R. V2V path loss modeling for example 5 GHz overpass channels. *79thVehicular Technology Conference (VTC Spring), May 18–21, 2014, Seoul, South Korea*. Piscataway, NJ: IEEE Press, 2014: 1–5.

[47] Liu P, Ai B, Matolak D W, et al. 5 GHz vehicle-to-vehicle channel characterization for example overpass channels. *IEEE Transactions on Vehicular Technology*, 2016, 65(8): 5862–5873.

[48] Pekka K, Meinilä J, Hentila L, et al. WINNER II channel models part II radio channel measurement and analysis results. Chichester: A John Wiley and Sons, Ltd., 2007.

[49] Parviainen R, Kyosti P, Hsieh Y, Ting P, Chiou J. Results of high speed train channel measurements: EURO-COST 2100. *European Cooperation in the Field of Scientific and Technical Research*, 2008, 1–6.

[50] Dong W, Liu G, Yu L, Ding H, Zhang J. Channel properties of indoor part for high-speed train based on wideband channel measurement. *5th International ICST Conference on Communications and Networking in China (CHINACOM), August 25–27, 2010, Beijing, China*. Piscataway, NJ: IEEE Press, 2010: 1–4.

[51] Liu L, Tao C, Qiu J H, et al. Position-based modeling for wireless channel on highspeed railway under a viaduct at 2.35 GHz. *IEEE Journal on Selected Areas in Communications*, 2012, 30(4): 834–845.

[52] Sun R, Tao C, Liu L, et al. Nonisotropic scattering characteristic in an alternant tree-blocked viaduct scenario on high-speed railway at 2.35 GHz. *International Journal of Antennas and Propagation*, 2014, 2014, Article ID 642894: 9.

[53] Zhou T, Tao C, Liu L, Tan Z. A semiempirical MIMO channel model in obstructed viaduct scenarios on high-speed railway. *International Journal of Antennas and Propagation*, 2014, 2014. Article ID 287159.

[54] Sun R, Tao C, Liu L, Tan Z. Channel measurement and characterization for HSR U-shape groove scenarios at 2.35 GHz. *2013 IEEE 78th Vehicular Technology Conference (VTC Fall), September 2–5, 2013, Las Vegas, NV, USA*. Piscataway, NJ: IEEE Press, 2013: 1–5.

[55] Zhou T, Tao C, Liu L, Tan Z. Ricean K-factor measurements and analysis for wideband radio channels in high-speed railway U-shape cutting scenarios. *IEEE Vehicular Technology Conference (VTC)-Spring, May 18–21, 2014, Seoul, South Korea*. Piscataway, NJ: IEEE Press, 2014: 1–5.

[56] Elektrobit. *PropSoundTM CS Multi-Dimensional Channel Sounder General Presentation*. Finland: Elektrobit, 2005.

CHAPTER 3

Wireless Channel Modeling Theory and Methods

T HE WIRELESS CHANNEL MODEL is an abstract description of channel characteristics established by people after sufficient research on the wireless propagation environment and its propagation characteristics. The wireless channel is the transmission medium of mobile communication, and all information is transmitted in this channel. The quality of channel performance directly determines the quality of communication. Therefore, to transmit as much useful information as possible on limited spectrum resources, we must have a very clear understanding of the characteristics of the channel to adopt corresponding technologies according to the characteristics of the channel. Fight against interference and fading, and ensure transmission quality and transmission capacity. Wireless channel models are widely used in the research of communication systems: when designing wireless transmission technology and optimizing mobile wireless communication systems, it is necessary to fully consider channel constraints. Design and optimization of mobile wireless communication systems that meet the needs; standardization organizations and various research institutions also need to use certain channel models and business models to evaluate and select mobile wireless communication system standards; System planning works to determine system parameters and ensure efficient use of frequency resources. At this time, a reasonable channel model can help designers to correctly estimate and select system parameters such as the location and coverage of each base station, antenna configuration and MIMO scheme, data rate during transmission, and system capacity, so that the results of the system planning have higher credibility. Only by establishing a real wireless channel model that conforms to the network deployment of the communication system and typical scenarios, and applying it in the design, performance evaluation and standardization of transmission technology and system, can the actually deployed system perform at its best and reduce the cost of optimization and maintenance. The wireless channel model can also be used for system equipment test selection and network planning

DOI: 10.1201/9781032669793-3

optimization, to maximize the simulation of the actual wireless propagation environment, and to accelerate the industrialization and application of future communication standards.

Looking back at the development of mobile communication systems, the research on wireless channels is accompanied by the development of each generation of systems. The channel characteristics studied and the channel model established in each period are directly related to the system performance that people are most concerned about at that time. Early wireless communication systems, such as the first-generation and the early days of the second-generation system, were mainly used for voice communication, with low requirements on data rate, and the most concerned was the coverage of the system. Channel modeling in this period mainly focused on the strength of the signal, and most of the models given were path loss models, typically Okumura-Hata [1] and Lee model [2], etc. Then, as the data transmission rate requirements increased, the bandwidth used by the system also increased, and people began to conduct in-depth research on the effectiveness of the system. The increase in bandwidth means that the time occupied by each symbol in the system transmission process is shortened. When the time occupied by a symbol is less than the relative time delay between multipath, the sampling of the current symbol at the receiving end will mix the delayed signal components of the previously sent symbols, resulting in serious inter-symbol interference. From the perspective of the frequency domain, even if the system bandwidth is greater than the coherent bandwidth of the channel, it will cause the wideband signal to experience frequency selective fading when passing through the channel. Accurately describing the multipath propagation characteristics of the channel is of great significance to the design and evaluation of anti-multipath fading techniques such as equalization or OFDM. Therefore, people have gradually increased their research on multipath delay related parameters, and at the same time began to study the Doppler parameters reflecting the time-varying channel. The model at this time is generally a tap delay line (Tap Delay Line, TDL) model; it contains parameters such as energy, delay, and Doppler of each propagation multipath. The M.1225 model [3] of ITU (International Telecommunication Union, International Telecommunication Union) and 3GPP (3rd Generation Partnership Project, Third Generation Partnership Project) spatial channel model (Spatial Channel Model, SCM), SCM extension model (Spatial Channel Model Extension, SCME) [4,5].

With the continuous development of communication systems, people are eager to further increase the transmission rate of the system. In addition to more in-depth development of the time domain and frequency domain resources of the channel, research on the air domain has also begun, especially the introduction of MIMO and smart antenna technology, channel measurement, and modeling also pay more attention to the spatial information of multipath. At the same time, as the bandwidth continues to increase and the resolution of multipath continues to improve, the characteristics of the delay domain are also changing. The channel model develops into a cluster delay line (Clustered Delay Line, CDL) model based on the TDL model. The difference between the CDL model and the TDL model is the introduction of the concept of clusters [6–8], that is, a collection of multipath with similar arrival angles, departure angles, and time delays.

The introduction of clusters reflects the characteristics of multipath in the channel in more detail, and at the same time introduces spatial information such as angle of arrival and angle of departure of multipath, which is very important for the research of MIMO technology and smart antenna technology. At present, the TGn model [9] and WINNER (Wireless World Initiative New Radio) of IEEE (Institute of Electrical and Electronics Engineers, Institute of Electrical and Electronics Engineers) 802.11 for WLAN (Wireless Local Area Network, wireless local area network) are more commonly used. Organize the CDL model [10] in different scenarios for the IMT-Advanced (International Mobile Telecommunications-Advanced) system.

Traditional modeling methods include statistical modeling; deterministic modeling methods include statistical, deterministic, and geometry-based stochastic channel modeling. Deterministic modeling is based on the basic mechanism of radio wave propagation. When the environmental parameters are fully grasped, accurate prediction of radio wave propagation can be achieved, but complex and time-consuming calculations have also become an insurmountable gap in deterministic modeling. In this case, statistical modeling can more easily and accurately reflect the basic characteristics of wireless channels, so the research results on this method are particularly fruitful, but the large test overhead and long test cycle also discourage many scholars. The stochastic modeling method is a combination of the above two methods, which is widely used in the 4G and 5G communication era. Next, we will discuss the above-mentioned channel modeling theories and methods in detail.

3.1 STATISTICAL CHANNEL MODELING

Statistical channel modeling refers to conducting transmission and reception experiments in various typical propagation environments and recording relevant data of received signals or channels on site. After the actual measurement, a large amount of data is statistically analyzed by computer to find the statistical laws of various parameters reflecting the channel transmission characteristics, and a propagation model is established according to the data analysis results. The statistical modeling method is simple and does not need to consider detailed environmental information during application, so it is easy to implement; but generally speaking, the accuracy of statistical modeling methods for channel parameter prediction is not high, and it cannot reflect the impact of subtle changes in the specific environment on the channel. Several typical statistical channel models will be introduced from different perspectives below.

3.1.1 Narrowband Model

When a single pulse signal reaches the receiving end after being transmitted through a multipath channel, due to the existence of various backscattering objects in the channel, the received signal will become a pulse train, and each pulse in the burst corresponds to a direct component and a multipath component that occurs at each time delay, which shows that it has a time-delay spread property. In practical applications, there are various parameters describing the delay spread, such as root mean square delay spread, average delay spread, etc. [11]. But the delay spread of the channel often changes with time and becomes a random variable. Therefore, the time delay characteristics of different channels are usually compared by using the statistical characteristic root mean square delay spread τ_{rms} obtained from the power delay spectrum. After the channel delay spread is quantized, the channel model can be divided into a narrowband channel model and a wideband channel model according to the relationship between the symbol width of the transmitted signal and the channel delay spread.

In the narrowband channel model, the transmitted signal symbol width is much larger than the delay spread of the channel, resulting in that although the arrival time of each multipath component at the receiving end is different, it is indistinguishable from the receiving end. The total received signal is formed after the end vectors are superimposed, as shown in Figure 3.1a. Since the multipath signals propagate along different propagation paths, there is a difference in the phase between each other. When the multipath components are synthesized at the receiving end, the total signal amplitude is constructive or destructive, so that the amplitude of the received signal presents random and rapid changes. In practice, the probability density distribution function of the envelope of the received signal is usually used to describe the small-scale variation of the received signal in a narrowband channel. The commonly used distribution models are the Ricean distribution and the Rayleigh distribution. At this time, $\tau = \tau_l$ in equation (3.1) can be set to obtain the narrowband time-invariant channel model equation (3.2), namely:

$$h_n(\tau) = \sum_{l=0}^{N-1} a_l \cdot e^{j\phi_l} \cdot \delta(\tau - \tau_l) \tag{3.1}$$

$$h_n(\tau) = \sum_{l=0}^{N-1} a_l \cdot e^{j\phi_l} \tag{3.2}$$

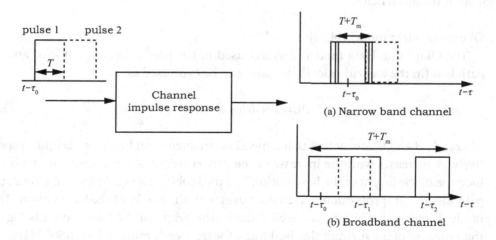

FIGURE 3.1   Schematic illustration of a pulse train passing through a multipath channel.

3.1.1.1 Small-Scale Fading and Large-Scale Fading Modeling

For a narrowband channel, the impulse response is formed by the decay of an impulse function over time, so for a slowly time-varying channel, it can be expressed as:

$$h(t,\tau) = \alpha(t)\delta(t) \tag{3.3}$$

Among them, the fluctuation of the amplitude factor $\alpha(t)$ in a small range is usually modeled as a random process, and its autocorrelation function is determined by the Doppler spectrum. The complex magnitude is modeled as a zero-mean, circularly symmetric complex Gaussian random variable. When the backscatterers around the mobile station are evenly distributed and there is no direct light, the absolute value of the amplitude obeys the Rayleigh distribution, which is also customarily called Rayleigh fading.

When considering the fluctuation of the signal in a larger range, the small-scale average amplitude F is subject to the logarithmic normal distribution, and the standard deviation is defined as σ_F; a large number of tests indicate that most of the empirical values of σ_F are 4~10 dB. The spatial autocorrelation function of lognormal shadow fading is usually assumed to be a bilateral exponential distribution, and the correlation distance ranges from 5 to 100 m in different scenarios.

3.1.1.2 Path Loss Model

The modeling of the average strength of the received signal (the received signal has been averaged to filter out large-scale fading and small-scale fading at the same time) is path loss modeling, and deterministic methods can also be used in specific scenarios. Among the statistical models, the simplest models are the free-space path loss model and the two-propagation "breakpoint" model (for distances $d < d_{\text{break}}$, $n = 2$; for larger distance ranges, $n = 4$). In practice, the path loss depends not only on the distance, but also on other external parameters, such as the height of the transceiver antenna, the height of buildings in the environment, and the real measurement environment. Some common path loss models [12,13] are introduced below.

1. Okumura–Hata model [14,15]:

    The Okumura–Hata model is widely used in the prediction of path loss, and its path loss (in the logarithmic dB domain) can be expressed as:

$$\text{PL} = A + B \log d + C \tag{3.4}$$

where, A, B and C are factors that depend on frequency and antenna height, respectively. A increases with the increase of the carrier frequency, and decreases with the increase of the height of the base station and the mobile station. At the same time, the path loss factor (proportional to B) decreases with the height of the base station. This model is only suitable for macrocells; that is, the height of the base station is higher than the roof of the surrounding buildings, the frequency range is 150~1,500 MHz, the base station antenna range is 30~200 m, and the mobile station height range is 1~10 m.

2. COST231 – Walfisch–Ikegami model [16]:

   The COST231 – Walfisch–Ikegami model is mainly suitable for microcells and smaller macrocells because it has less restrictions on the distance between the base station and the mobile station and the height of the antenna. In this model, the total path loss is modeled by the free-space path loss L_0, the multi-frequency loss along the propagation path, and the attenuation L_{msd} (roof-to-street diffraction and scattering loss) from the last roof edge to the mobile station L_{rts}, as shown in Figure 3.2, it can be specifically expressed as:

$$L_b = \begin{cases} L_0 + L_{rts} + L_{msd}, & L_{rts} + L_{msd} > 0 \\ L_0 & , L_{rts} + L_{msd} \le 0 \end{cases}$$

(3.5)

The free space loss depends on the carrier frequency and distance, while the roof-to-block diffraction loss depends on the frequency, the width of the street, the height of the mobile station, and the angle between the street direction and the base station-mobile station connection direction. The multi-frequency loss depends on the distance between buildings, the distance between the base station and the mobile station, the carrier frequency, the height of the base station, and the height of the roof. The model assumes a Manhattan street pattern (street intersections at right angles), constant building heights, and flat terrain. In addition, the model does not include the effect of waveguide propagating through street canyons, which can affect the received signal strength.

3. Motley–Keenan model:

   For interior scenes, the attrition of walls plays an important role. Based on this consideration, the path loss in the Motley–Keenan model can be expressed as:

$$PL = PL_0 + 10n\log\left(\frac{d}{d_0}\right) + F_{wall} + F_{floor}$$

(3.6)

where F_{wall} is the sum of the losses caused by the multipath components propagating from the transmitter to the receiver through the wall; similarly, F_{floor} describes the sum of the losses caused by the floor between the transmitter and the receiver.

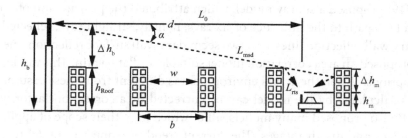

FIGURE 3.2   COST231 – Parameters in the Walfisch–Ikegami model.

According to different building materials, at 300 MHz~5 GHz, the loss caused by a wall is 1~20 dB, and the loss caused by a wall in the high frequency band is even higher. The Motley–Keenan model is a scene-dependent model that needs to provide the specific location of the transmitter and receiver and the structure of the building. However, this model is not completely accurate because it ignores multipath that propagates "around" the wall. For example, a wave traveling between two distant offices can penetrate many walls (collimated beam paths) or through a corridor (the signal leaves one of the offices, propagates through the corridor, and eventually reaches the receiver in the other office). The latter mode of propagation is usually more efficient, but is not considered in the Motley–Keenan model.

4. Double slope model [17]:

The double-slope model is also called the wideband PCS microcellular model. Feuerstein et al. [18] in San Francisco (San Francisco, USA) and Oakland (Oakland, USA) in 1991, using a 20 MHz pulsed transmitter on 900 MHz band, tested for path loss, outage rate and delay spread in LOS and obstructed environments. This model can be used to estimate the LOS microcell path loss. The path loss model expression in the LOS situation is as follows (equation (3.7)):

$$L(d) = L_b + \begin{cases} 10n_1 \log d + P_1 & ,1 < d < d_{brk} \\ 10(n_1 - n_2)\log d_{brk} + 10n_2 \log d + P_1 ,d \geq d_{brk} \end{cases} \tag{3.7}$$

Among them, L_b is the base station transmission loss, which is related to the frequency and the antenna height of the transmitting station. When there is an obstacle, equation (3.7) is simplified to: $L(d) = 10n \log d + P_1$, which is a typical logarithmic distance path loss model. P_1 is the signal power at the reference distance. In large-coverage cellular systems, the reference distance is often 1 km, and in microcells, a smaller distance (1 or 100 m) is often used. The reference distance should always be set at the far field of the antenna to avoid the near-far effect on the reference path loss.

5. Other models:

In addition to the traditional models mentioned above, many experts and scholars have also proposed other empirical and theoretical models. For example, Rustako et al. [19] proposed a six-ray model, which attributed the propagation of radio waves with a LOS path to the influence of six rays, namely, straight rays, reflected rays, two primary wall reflection lines, and two secondary wall surface reflection lines. Lee [20] also proposed an area correlation model suitable for flat terrain. This model has many assumptions. When the actual environment is different from these assumptions, the results calculated by the model can be corrected by a correction factor. References [21–24] also proposed many models, all of which have their scope of application and advantages and disadvantages. The current trend of empirical model research is to combine the existing large-scale models to improve the prediction accuracy of the combined model. Reference [25] combined the three models of COST231, Walfisch-Ikegami Flat Edge, and Maciel Xia Bertoni to obtain a more accurate path loss prediction model.

There are two main differences between the indoor propagation environment and the outdoor propagation environment: the coverage distance becomes smaller, and the environmental change becomes more significant for a smaller distance between the transmitter and receiver [13]. The propagation in the building is affected by the structure of the building the influence of factors such as layout, material and building type. References [26–30] present models suitable for indoor radio wave prediction. The main indoor scene model is the attenuation factor model Seidel [31], which describes the changes in propagation characteristics affected by building types and occlusions. This model has been used to accurately calculate the wireless coverage of indoor and campus networks. The model is highly flexible and reduces the standard deviation of predicted path loss to reagent measurements to 4 dB or so. The mathematical expression of the model is as follows:

$$PL(d)[dB] = PL(d_0)[dB] + 10n_{SF} \log\left(\frac{d}{d_0}\right) + FAF[dB] + \sum PAF[dB] \qquad (3.8)$$

Among them, n_{SF} represents the index value of the same floor test, FAF represents the attenuation factor of a specific number of floors of the building, and PAF represents the division attenuation factor of a specific obstacle encountered by the ray propagation between the transmitter and receiver in three-dimensional space. If the attenuation factor of the floor is not considered, the model can be expressed as:

$$PL(d)[dB] = PL(d_0)[dB] + 10n_{MF} \log\left(\frac{d}{d_0}\right) + \sum PAF[dB] \qquad (3.9)$$

Devasirvatham et al. [18] found that the path loss in the building is equal to the free space loss plus the additional loss factor, and it increases exponentially with the increase of the distance. The mathematical expression of the revised model is as follows:

$$PL(d)[dB] = PL(d_0)[dB] + 20 \log\left(\frac{d}{d_0}\right) + \alpha d + FAF[dB] + \sum PAF[dB] \qquad (3.10)$$

3.1.2 Wideband Model

In the wideband channel model, the delay spread of the channel is relatively large, so that all or part of the multipath components of the same transmitted signal can be distinguished in the delay domain at the receiving end. The received signal at this time can be expressed as a series of discrete pulses with a time delay of τ_l. If the transmitter sends signals continuously, due to the extension of the duration of the previous information bit, its multipath component may arrive in the time slot of the next information bit, resulting in Inter-Symbol Interference (ISI), as shown in Figure 3.1b. Common wideband channel models are as follows.

3.1.2.1 Multipath Delay Model

The most widely used wideband channel model is the N-path Rayleigh fading channel model. This is a relatively common structure, and it is essentially a TDL structure, but with the restriction on the amplitude, that is, the amplitude of all taps obeys the Rayleigh fading distribution, and an additional direct path is added. The channel impulse response of this model can be expressed as:

$$h(t,\tau)=a_0\delta(\tau-\tau_0)+\sum_{i=1}^{N}c_i(t)\delta(\tau-\tau_i) \tag{3.11}$$

Among them, the amplitude parameter of the direct beam a_0 does not change with time, but the amplitude $c_i(t)$ is a complex Gaussian random process with zero mean value, and its autocorrelation function depends on the corresponding Doppler spectrum (such as Jakes spectrum). In most cases, $\tau_0=\tau_1$, so the magnitude of the first tap obeys the Rice distribution. The model can be further simplified when the number of taps is limited to $N=2$, and it can be assumed that there are no direct paths. This is the simplest random fading model that embodies channel delay dispersion, so it is widely used in theoretical analysis.

Another commonly used channel model consists of a direct path component determined by one hop and a fading tap ($N=1$), and its delay τ_0 can be different from τ_1. This model is widely applicable to channels such as satellite communication channels, where direct links usually exist and paths reflected by buildings close to the receiving end produce a time-delayed fading component. At that time $\tau_0=\tau_1$, this channel was simplified as a Rice channel with flat fading.

3.1.2.2 Multipath Power Spectrum Model

A large number of channel measurements show that the multipath power delay profile (Power Delay Profile, PDP) can be approximated as a unilateral exponential function:

$$P_h(\tau)=P_{SC}(\tau)=\begin{cases}\exp\left(-\dfrac{\tau}{S_\tau}\right)\tau\geq 0\\[2mm]0,\ \text{Others}\end{cases} \tag{3.12}$$

In a more general model, the PDP consists of the sum of multiple time-delayed exponential functions, equivalent to a multi-cluster interactor, namely:

$$P_h(\tau)=\sum_l\frac{P_l^c}{S_{\tau,l}^c}P_{SC}\left(\tau-\tau_{0,l}^c\right) \tag{3.13}$$

Among them, P_l^c, $\tau_{0,l}^c$, and $S_{\tau,l}^c$ are respectively the power, delay, and delay spread of the lth cluster. The sum of the powers of all clusters is the narrowband power. For the PDP form in equation (3.12), the RMS delay spread characterizes the delay spread. Most of the measurements in the existing literature still use this parameter to characterize the delay spread. Typical values of delay spread in different environments are as follows.

- Indoor residential buildings: the typical value is 5~10 ns, but the maximum value that has appeared in the measurement is 30 ns.

- Indoor office scene: The typical value of the delay spread in this scene is 10~100 ns, but the maximum value that has appeared in the measurement is 300 ns. The size of the room has a significant impact on the value of the delay spread, and the size and shape of the building also have varying degrees of influence on the delay spread.

- Factories and airport halls: The typical value of the delay spread in this scenario is 50–200 ns.

- Microcell: In a microcell, the typical value of the delay spread is 5~100 ns (for the direct shot condition) and 100~500 ns (for the non-direct shot condition).

- Tunnels and pits: Empty tunnels usually have a small delay spread (around 20 ns), while tunnels with many vehicles usually have a large delay spread (up to 100 ns).

- Typical urban and suburban environments: Typical values for delay spread are 100–800 ns, and μs values as high as 3 have been observed.

- Urban and mountainous areas with poor environments: This type of scene is a typical multi-cluster scene, so it has a large delay extension. In measurements in European cities, values of delay spread of up to 18 μs and cluster delays of up to 50 μs were measured, while in US cities the values were lower. The cluster delay can reach 100 μs in mountainous scenes.

The delay spread is a function of the base-to-mobile separation, increasing approximately with distance, i.e., d^α, among urban and suburban scenarios $\alpha = 0.5$, in mountainous areas $\alpha = 1$. Delay spread also typically exhibits large large-scale fluctuations. Many references find that delay spread is log-normally distributed in urban and suburban scenarios with a standard deviation of 2~3 dB.

3.1.2.3 Arrival Time Model of Multipath and Multipath Clusters

For systems with higher bandwidths, the resolution of the multipath components is higher. In this case, it is more effective to describe the PDP with the arrival time of the multipath time delay, and add an envelope equation to describe the power value of the multipath component concerning the time delay. To statistically model the time-of-arrival of multipath components, a first-order approximation is to assume that objects causing reflections in an urban environment are randomly distributed in space such that the excess delay follows a Poisson distribution. However, measured data show that multipath components usually arrive in groups, i.e., clustering characteristics. Two types of models are proposed to reflect this fact: $\Delta - K$ the model and the Saleh–Valenzuela (SV) model, as follows.

- $\Delta - K$ model. The model has two states: S_1, the average arrival rate of the multipath signal is $\lambda_0(t)$; S_2, the average arrival rate is $K\lambda_0(t)$. The process of multipath arrival begins with the state S_1. Transition to state S_2 at time interval $[t, t + \Delta]$ if the arrival time of a multipath component is t. Return to state at the end of the time interval if no other multipath arrive during the time interval S_1. It should be pointed out that for $K = 0$ or $\Delta = 1$, the above process is simplified to a standard Poisson process.

- SV model [32]. The model assumes the existence of multipath clusters a priori. Within a cluster, the arrival of multipath components obeys the Poisson distribution, and the arrival time of the cluster itself also obeys the Poisson distribution (only the time interval constant is different). In addition, the power of multipath components within a cluster decreases exponentially with delay, and the power of a cluster follows another (different) exponential distribution, as shown in Figure 3.3.

Mathematically, SV models usually use the discrete-time impulse response shown in equation (3.14):

$$h(\tau) = \sum_{l=0}^{L} \sum_{k=0}^{K} c_{k,l}(\tau) \delta(\tau - T_l - \tau_{k,l}) \tag{3.14}$$

Among them, the distribution of cluster arrival time and path arrival time is described as:

$$\text{pdf}(T_l | T_{l-1}) = \Lambda \exp[-\Lambda(T_l - T_{l-1})], \, l > 0 \tag{3.15}$$

$$\text{pdf}(\tau_{k,l} | \tau_{(k-1),l}) = \lambda \exp\left[-\lambda(\tau_{k,l} - \tau_{(k-1),l})\right], \, k > 0 \tag{3.16}$$

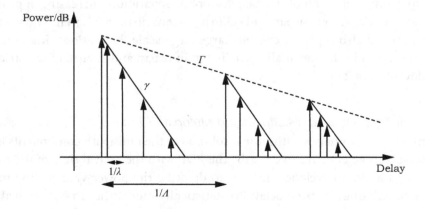

FIGURE 3.3 SV model schematic.

Among them, T_l represents the time of the first arrival path of the lth cluster, $\tau_{k,l}$ is the time delay of *the kth path* in the lth cluster relative to the first arrival path (defined as: $\tau_{0,l} = 0$), Λ is the cluster arrival rate, and λ is the path arrival rate, i.e., the arrival rate of each cluster inner diameter. The dependence of all the above data on absolute time has been eliminated. The PDP within each cluster can be expressed as:

$$E\left\{|c_{k,l}|^2\right\} \propto P_l^c \exp\left(\frac{-\tau_{k,l}}{\gamma}\right) \tag{3.17}$$

Among them, P_l^c is the energy of the lth cluster, γ and is the decay time constant in the cluster. The power of the cluster exhibits an exponential decay form, namely:

$$P_l^c \propto \exp\left(\frac{-T_l}{\Gamma}\right) \tag{3.18}$$

3.1.2.4 Standardized Model

A special case of the TDL model is the COST207 model, which defines the channel PDP or tap weights and Doppler spectra in four typical environments. These PDPs are distilled from a large number of measurements in Europe – i.e., typical urban areas, poor urban areas, rural areas, and mountainous areas. Depending on the scenario, the PDP can be expressed as a single exponential decay or two single exponential equations with a time delay between them. The second cluster corresponds to distant high-rise buildings or mountains that are equivalent to effective interactors, thus producing a set of multipath components with considerable power and time delay.

The COST207 model is a channel model developed based on measurement data with a very narrow bandwidth, and is only applicable to systems with a bandwidth of 200 kHz or less. For the simulation of the third-generation cellular system with a bandwidth of 5 MHz, the International Telecommunication Union (ITU) proposed another series of models to consider the large bandwidth system, which is divided into pedestrian, vehicle, and indoor scenarios [12].

3.1.3 Directional Channel Model

The directional channel model mainly originated from the application of MIMO technology in the field of wireless channel measurement and modeling. Compared with the traditional channel model, the directional channel model mainly adds the description of the angle domain characteristics of the multipath component, realizes the separation of the channel model and the antenna model, and has better universality [12,33], as follows.

3.1.3.1 General Model Structure and Factors

A relatively general model is usually based on the Double Directional Delay-Power Spectrum (DDDPS), which depends on three variables: the angle of departure (Direction of Departure, DOD), the angle of arrival (Direction of Arrival, DOA), and latency.

An important simplification can be obtained if DDDPS can be decomposed into three equations, where each equation depends on only one parameter, namely:

$$\text{DDDPS}(\Omega,\Psi,\tau)= \text{APS}^{\text{BS}}(\Omega)\text{APS}^{\text{BS}}(\Psi)P_h(\tau) \tag{3.19}$$

This shows that the angular power spectrum (APS) on the base station side is independent of delay, and the mobile APS is also independent of delay. In addition, the APS of the mobile terminal is independent of the direction of the antenna pattern of the base station transmitting terminal, and vice versa.

This theoretical decomposition greatly simplifies the theoretical computational complexity as well as the parameterization of the channel model. However, this approach does not always reflect physical reality. A more general model assumes that DDDPS consists of multiple clusters; each cluster has an independent DDDPS, namely:

$$\text{DDDPS}(\Omega,\Psi,\tau)= \sum_l P_l^c \text{APS}_l^{c,\text{BS}(\Omega)\text{APS}^{c,\text{MS}}(\Psi)_{h,l}^c(\tau)} \tag{3.20}$$

Among them, the superscript c indicates the cluster, and l indicates the number of the cluster. Obviously, if only one cluster exists, this model will reduce to the model in equation (3.18).

3.1.3.2 Angular Dispersion on the Base Station Side

The most common model of APS on the base station side is the Laplace distribution in the horizontal angle direction:

$$\text{APS}(\phi)\propto \exp\left[-\sqrt{2}\,\frac{|\phi-\phi_0|}{S_\phi}\right] \tag{3.21}$$

Where ϕ_0 is the mean value of the horizontal angle. The pitch angle spectrum is usually modeled as an impulse function (i.e., all emitted rays are incident on the horizontal plane), therefore, $\Omega=\phi$. The spectrum can also be modeled as a Laplace distribution.

Below are some typical ranges of values for RMS Angle Spread and Cluster Angle Spread.

- Indoor office scene: Under non-direct sunlight conditions, the root-mean-square angle spread of the clusters is $10°\sim20°$, while under direct sunlight conditions, the value is approximately $5°$.

- Industrial scene: Under non-direct sunlight conditions, the root mean square angle expands to $20°\sim30°$.

- Microcell: The RMS angle spread is $5°\sim$ in the direct beam condition and $20°\sim40°$ in the direct beam condition $10°$.

- Typical urban and suburban scenes: In the dense urban scene, the measured root mean square angle spreads to $3°\sim20°$; in the suburban scene, the root mean square angle spreads smaller due to the frequent birth and death of direct paths.

- Poor urban and mountainous scenes: Due to the existence of multipath clusters, the root mean square angle is expanded to be 20° or larger.

- Rural scene: RMS angle expanded to 1° ~ 5°.

In outdoor scenarios, the angular spread distribution in a larger range also obeys the lognormal distribution and is correlated with the angular delay spread. This allows the extended logarithm to be viewed as a correlated Gaussian random variable. In addition, a large number of practical tests have shown that there is also a correlation between the channel angle spread and the distance between the transceiver and the transmitter.

3.1.3.3 Angle Dispersion of the Mobile Terminal

For outdoor scenes, since the mobile station is surrounded by local cars, people, houses, etc., it is usually assumed that rays arrive at the mobile station from all directions. This model dates back to the 1970s. However, recent studies have shown that the horizontal angular spread can be very small, especially in street canyons. The APS can therefore still be approximated as a Laplace distribution; the angular spread of the clusters is about in the order of 20°. Furthermore, angular distribution is an equation for the time delay. For the case where the mobile station is located in the canyon of the street and has no direct path, the path propagating above the roof has a small delay but also produces a large delay spread, while the subsequent multipath propagates along the street and thus has a small Angle extension. Under indoor (quasi-)direct illumination conditions, paths arriving first have a smaller angular spread, while paths with larger delays have uniformly distributed APS.

For the outdoor pitch angle spectrum, the pitch angle distribution of the multipath propagating above the roof is a uniform distribution between 0 and the angle at which the roof can be seen; the pitch angle of the path arriving along the street canyon follows the Laplace distribution.

3.1.3.4 Polarization Characteristics

Most of the channel models only analyze the vertical polarization in the propagation process, which means that both the transmitting end and the receiving end use vertically polarized antennas. However, interest in polarization diversity research continues to grow. To simulate the system, a propagation model with dual-polarized radiation characteristics is required.

A signal transmitted from a vertically polarized antenna undergoes some interactions before reaching the receiver, resulting in energy conversion to the horizontally polarized direction (and vice versa). Therefore, the fading coefficient of the multipath component needs to be expressed as a polarized 2×2 matrix, and the complex magnitude a_l becomes:

$$a_l = \begin{pmatrix} a_l^{VV} & a_l^{VH} \\ a_l^{HV} & a_l^{HH} \end{pmatrix} \tag{3.22}$$

Where V and H represent vertical and horizontal polarization, respectively.

The most common polarized channel models assume that the inputs in the matrix are all statistically independent and complex Gaussian fading random variables. It is assumed that the mean powers of the sum components are independent of each other; similarly, VH the HV mean powers of the sum components are the same. Cross-Polarization Discrimination (XPD) is the VV ratio VH of the average power of the sum component (in the dB domain) and is modeled as a Gaussian random variable. The mean and variance of XPD depend on the propagation environment and even on the time delay of the considered multipath. The typical value of XPD mean is 0~12 dB, while the standard deviation is 3~6 dB.

3.1.3.5 Model Realization

The continuous model concerning the angle spectrum was discussed above. For computational implementations, a discrete version is usually required. One way to implement a directional channel model is the generic TDL model, in which DDDPS is discrete. Another possible model is the geometrically based stochastic channel model (GSCM). In this approach, instead of random modeling of the strength and angle of the multipath components, the position of the interacting bodies and the strength of the interaction are random (as shown in Figure 3.4). In addition, it is necessary to assume the existence of only one interaction process. The direction-resolvable impulse response can be obtained through the following two steps [12].

Step 1 According to the probability density function of the position of the interacting body, it is assigned position coordinates.

Step 2 determines the contribution of the interactors to the bidirectional impulse response based on the assumption that there is only one interaction process. Each multipath component (corresponding to an interactor) has a unique DOA, DOD, amplitude delay, and phase offset.

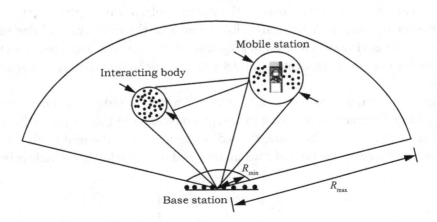

FIGURE 3.4   Schematic diagram of the realization of the random channel model based on geometry.

The simplest model assumes that all interacting bodies are close to the mobile. This situation usually occurs in macrocells with regular building structures, such as suburban scenes. In this case, rays from the mobile station interact with interacting volumes around it, but not necessarily with interacting volumes around the base station. Different models exist for the distribution of interacting bodies around the mobile station.

- Some studies place all interacting bodies in a circle around the mobile station.

- There are also some studies that suggest an even distribution within a disc. When the mobile station moves, the disk around the mobile station also moves [34]. Some interactors will move out of the disc, and some new interactors will enter the disc region (as shown in Figure 3.5). This also corresponds to the physical fact that the interacting body of the principal mobile does not make a significant contribution (although physically it still exists).

- Other studies suggest using a unilateral Gaussian distribution function, namely: $\mathrm{pdf}(r) = \exp\left(\dfrac{-r^2}{2\sigma^2}\right), r \geq 0$. Calculating PDP and APS through this distribution gives the results shown in Figures 3.6 and 3.7 [35]. It can be seen that these results are quite similar to the exponential PDP and the Laplacian APS.

When all interacting bodies are near the mobile station, this situation is the "single cluster" situation, and its angular delay power spectrum (Angular Delay Power Spectrum, ADPS) is approximated as:

$$\mathrm{ADPS}(\tau,\phi) \propto \exp\left(\frac{-\tau}{S_\tau}\right)\exp\left(\frac{-\sqrt{2}\,|\phi-\phi_0|}{S_\phi}\right) \qquad (3.23)$$

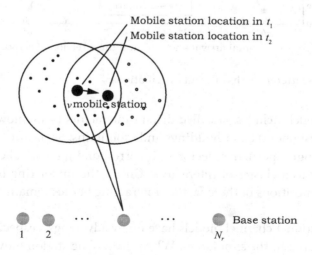

FIGURE 3.5 Interactors "disappear" or "appear" as the mobile moves (assuming all interactors are in a disk around the mobile. Scatterers only act at time instants t_1 (t_2) marked black (empty) circle; those that have an effect at both moments are marked as gray circles).

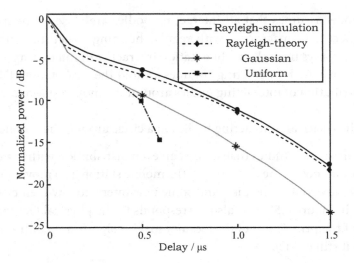

FIGURE 3.6   PDP of different distributions of scatterers.

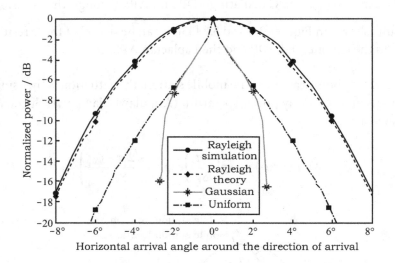

FIGURE 3.7   APS of scatterers with different distributions.

The generalized model includes so-called distant interactors (also known as distant scatterers), which correspond to taller buildings and mountains. Such a distant interactor can be modeled as a single specular reflector (i.e., corresponding to a taller building with a smooth surface) or as a cluster of interactors. Unlike the interacting bodies around the mobile station, the positions of these farther interacting bodies remain constant throughout the simulation.

Geometric directional channel models have many advantages, especially when motion needs to be considered in the simulation. When the mobile station moves, the parameter adjustment of multipath components can be done automatically. Therefore, the correct fading correlation results can be automatically obtained from the motion; meanwhile, the correlation characteristics of the mobile terminal DOA and Doppler frequency shift are also considered. Any variation in average DOA, DOD, and delay due to the large-scale motion of the mobile station is taken into account, which is difficult to implement in a tapped delay line model.

3.1.3.6 Standardized Directional Model

The European research group COST259 proposed a directional channel model, which has been widely recognized. The model is very realistic and includes a large number of effects and their interactions in different scenarios. Because the model covers a wide range of topics, this section only summarizes some of its basic properties. The COST259 directional channel model includes the large-scale and small-scale variation characteristics of the channel. It can be effectively realized through the following three levels.

- At the top level, different propagation scenarios (REs), i.e., classes of environments with similar propagation characteristics, are distinguished. There are 13 types of REs: four types of macrocell REs (base station height is higher than the roof), four types of microcell REs (outdoor, base station height is lower than the roof), and five types of picocell REs (indoor).

- The large-scale properties are described by their probability density functions, and the parameters vary from scene to scene. For example, delay spread, angle spread, shadow fading, Ricean factor, etc. all change with the movement of the mobile station in a large distance range. The realization of each large-scale fading parameter determines a DDDPS.

- At the third level, DDDPS is realized by a bidirectional impulse response and generated by small-scale fading.

Large-scale properties are described by a hybrid approach based on geometry and randomness, and this concept is applied to clusters of interacting volumes. At the beginning of the simulation, the clusters of interactors (one mobile station's local cluster and multiple clusters of distant interactors) are distributed randomly within the coverage area, which is the stochastic part. During the simulation, both the delay and the angle are determined by their position and the position of the base station and the mobile station, which is part of the geometry. Each cluster has a small-scale average DDDPS, its delay obeys the exponential distribution, the horizontal angle and the pitch angle of the base station side obey the Laplace distribution, and the horizontal angle and the pitch angle of the mobile station side obey the uniform or Laplace distribution distributed. The bidirectional impulse response can be directly obtained by averaging the ADPS, or it can be mapped to the distribution of the interacting body and then obtained geometrically.

The distribution of clusters in a macrocell is random. In microcells and picocells, these locations are determined using the concept of Virtual Cell Spread Area (VCDA). VCDA is a virtual town or office building, and has the mobile path of the mobile station. This method is similar to ray tracing, but has two important differences:

- "City map" does not need to reflect the actual city, so it can be used by many cities as a general model;

- Only the positions of the clusters are determined by ray tracing, while the behavior within a cluster is still randomly generated.

3.1.3.7 MIMO Matrix Model

The channel model with multipath component directional information was described above. Another very interesting concept in multi-antenna systems is to stochastically model the channel's impulse response matrix for a MIMO system. In this case, the channel is not only described by the statistical properties of the magnitude of each matrix input (typically Rayleigh or Ricean), but also the correlation between each input. The definition of the correlation matrix is: first "stack" all the inputs of the channel matrix into a vector $\boldsymbol{h}_{\text{stack}} = \left[h_{1,1}, h_{2,1}, \cdots, h_{N_r,1}, h_{1,2}, \cdots, h_{N_r,N_r} \right]^T$, and then calculate the correlation matrix $\boldsymbol{R} = E\left\{ \boldsymbol{h}_{\text{stack}} \boldsymbol{h}_{\text{stack}}^H \right\}$, where H is the Hermitian transformation. A popular simplified model assumes that the correlation matrix can be written as the Kronecker inner product $\boldsymbol{R} = \boldsymbol{R}_{\text{TX}}^T \otimes \boldsymbol{R}_{\text{RX}}$, where $\boldsymbol{R}_{\text{TX}} = E\left\{ \boldsymbol{H}^T \boldsymbol{H}^* \right\}$. $\boldsymbol{R}_{\text{RX}} = E\left\{ \boldsymbol{H} \boldsymbol{H}^H \right\}$ This model implies that the correlation matrix at the receiver is independent of the direction of transmission [36]; this is equivalent to assuming that the DDDPS can be decomposed into APSs that are independent at the transmitter and receiver. In this case, the channel transfer function can be written as:

$$ \boldsymbol{H} = \frac{1}{E\left\{ \text{tr}\left(\boldsymbol{H} \boldsymbol{H}^H \right) \right\}} \boldsymbol{R}_{\text{RX}}^{1/2} \boldsymbol{G}_G \boldsymbol{R}_{\text{TX}}^{1/2} \tag{3.24} $$

where $\boldsymbol{G}_G$ is a matrix with independent and identically distributed complex Gaussian inputs.

3.2 DETERMINISTIC CHANNEL MODELING

The deterministic modeling method is based on the electromagnetic wave propagation theory, that is, the path loss and other parameters reflecting channel characteristics are calculated by using the analysis method of electromagnetic wave propagation mechanism and theories of direct radiation, reflection, diffraction, and scattering, and the parameters of each test point of each path are obtained. In principle, the propagation channel of wireless communication can be regarded as a deterministic channel. Maxwell's equations together with the electromagnetic wave boundary conditions (position, shape, dielectric, and conductive properties of all objects in the environment) determine the field strength at all points and at all times. For outdoor scenes, such a deterministic model needs to consider all the geographical and morphological properties of the propagation environment; for indoor scenes, it needs to consider the characteristics of building structures, walls, and even furniture. Deterministic modeling methods do not rely on measurement results, but use a large number of terrain and building details to accurately predict the propagation loss and field strength distribution of electromagnetic waves. In this chapter, the basic principles of channel modeling will be introduced from a deterministic point of view.

To make deterministic modeling feasible, there are two main challenges [12]:

- a lot of computing time;
- Accurate understanding of boundary conditions.

Until about 1990, the approach to deterministic modeling was still limited by computation time and storage, but that has changed since then. On the one hand, computers have become so fast that what was difficult in 1990 even with a supercomputer can now be done with a personal computer. On the other hand, the development of more effective deterministic algorithms has also improved certain conditions.

Accurate boundary conditions are necessary for the successful application of deterministic models. This means that the location and electromagnetic properties of the entire "relevant" environment need to be known. The development of electronic regional maps and urban planning based on satellite photos and urban planning has also made great progress in recent years. The most accurate solutions (given a known environmental database) are those obtained by Maxwell's equations, requiring the use of integral or differential equations. The integral equations are mostly variants of the well-known method of moments, in which the induced currents in the interacting volume can be expressed by a series of fundamental equations. The simplest and most fundamental of these is the rectangular equation, extending over a fraction of a wavelength. Differential equations include the Finite Element Method (FEM) and the increasingly common Finite-Difference Time-Domain (FDTD).

All of these methods are very accurate, but the computational requirements also make them unusable in most scenarios. It is therefore more common to use approximations to Maxwell's equations as fundamental solutions. The most widely used approximation is the high-frequency approximation (also known as the ray approximation). In this approximation, electromagnetic waves are modeled as rays obeying the principles of geometric optics (Snell's law of reflection and transmission). In addition, diffraction and scattering can also take this approximation.

Most current deterministic models are based on electromagnetic methods of ray tracing, such as geometrical optics, physical optics, geometrical theory of diffraction (GTD), and uniform theory of diffraction (Uniform Geometrical Theory, UTD). Among the more classic models of this type are the Longly-Rice model [37–40], Durkin model [20,41], TIREM model [42,43], SEKE model [44,45] · Johnson-Gierhart [46–48] model, etc. These models are generally used to predict the attenuation characteristics of long-distance radio waves over irregular terrain, and belong to the large-scale path loss prediction method [49]. When applied, these models must consider the terrain parameters on the earth's great circle path between the transmitter and receiver, and consider the reflection, diffraction, refraction, and scattering of radio wave propagation by establishing various sub-model mechanisms. For example, the classic Longly-Rice model uses the two-ray interference model in geometric optics to predict the propagation characteristics of radio waves within the LOS in the environment, and uses the Fresnel-Kirchhoff single-edged peak model [13] to predict the diffraction loss of radio waves caused by isolated obstacles. The tropospheric scatter propagation of radio waves is predicted by the directional scattering theory, and the far-field diffraction loss of radio waves on the double-horizon path is predicted by the improved Van der Pol-Bremmer method [50]. Both the Durkin model and the TIREM model are very similar to the Longly-Rice model. When calculating the radio wave propagation loss in the LOS, the TIREM model chooses the one with the smaller diffraction loss on the smooth ground and the smaller loss in the reflection area. Path clearance judgment: The SEKE model mainly judges the multi-path interference area, multi-peak diffraction area, and smooth surface diffraction area of the radio wave through the clearance size on the propagation path, and establishes the corresponding propagation mechanism sub-model for these areas, and then finally uses the weighting method to calculate the radio wave propagation attenuation characteristics on the whole path. It can be seen that these large-scale path radio wave propagation models based on ray tracing methods not only need to establish equivalent sub-models reflecting each propagation mechanism according to specific environmental data, but also set the criteria of each sub-model, so they are relatively complicated. The amount of calculation is also very large. In addition, these models use statistical methods to approximate the complex atmospheric structure and complex surface electromagnetic characteristics in the propagation area. Therefore, the final prediction results are also statistical and do not have real-time prediction functions. 3D ray tracing methods based on GTD and UTD have been extensively studied and applied when predicting radio wave propagation characteristics on small-scale paths such as indoor or urban areas [51–54]. The ray tracing method is based on the "locality" of the high-frequency field, and simplifies the propagation of electromagnetic waves on the path into direct reflection and diffraction, so that the main propagation path can be searched according to the terrain environment database, and then according to the impact of each path on the field, the contribution ultimately yields the total field strength. Since the ray tracing method searches for the main propagation path according to the positions of the faces, splits, and vertices of the terrain, when the number of faces, splits, and vertices on the irregular terrain or the surface of the building is huge, the number of rays to be traced is very large, and the calculation is very complicated. In addition, some small-scale radio wave propagation models are not commonly used, such as the integral equation (Integral Equation, IE) method [55–59], the finite difference time domain (Finite-Difference Time-Domain, FDTD) method [60], iterative invariance test equation method (Measured Equation of Invariance, MEI) [61–63], and waveguide model [64,65]. Among them, due to the large amount of calculation, the FDTD method can only calculate the propagation characteristics of simple-structured cells or indoors. Although the calculation amount of the MEI method is small, it can only perform two-dimensional predictions at present, and the calculation of actual cells cannot obtain accurate results. Further details are given below.

3.2.1 Ray Emission

In the beam-emitting method, the transmitting antenna transmits the beams in different directions. Usually, the total spatial angle 4π can be divided into N units of the same size, and each ray is emitted to the center of one of the units [66] (i.e., the average sampling of

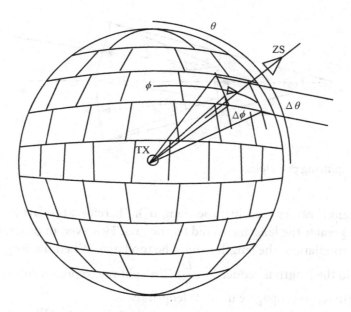

FIGURE 3.8 Principle of ray emission [12].

the spatial angle), as shown in Figure 3.8. The number of emitted rays is a trade-off between method accuracy and computation time.

The algorithm follows each propagating ray until it reaches the receiver or becomes so faint that it can be ignored (e.g., falls below the noise power). When following rays, there are some effects to consider.

- Free space loss. Because each path represents a specific angle in space, the energy per unit area d^{-2} falls in accordance with the propagation path of the ray.

- Reflections change the direction of ray propagation and cause additional losses. The reflection coefficient can be calculated by Snell's law, which is related to the angle of incidence and possibly to the polarization of the incident ray.

- Diffraction and scattering are taken into account by more advanced models. In these cases, the ray's incident on the interacting volume generates more new rays. The magnitude of the diffraction path is usually calculated by geometry or uniform diffraction theory.

The ray splitting algorithm has an important improvement in the accuracy of the method. The algorithm assumes that the effective intersection area of rays does not exceed a certain size (such as the size of a typical interacting volume). Therefore, if a ray is too far away from the transmitter after propagation, it will split into two rays. The principle can be further explained by Figure 3.9 [12]. To simplify the discussion, only the two-dimensional case is considered. Each ray represents not just a specific angle, but a range of angles with a width ϕ – corresponding to the angle between the two emitted rays. The length of the intersection of this angular region with a circle of radius d is approximately ϕd (for 3D, consider the

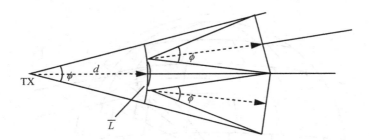

FIGURE 3.9   Ray splitting principle.

"intersection region" rather than just the "length"). Therefore, the farther away from the transmitter, the greater the length covered by the rays. However, in order to maintain the accuracy of the simulation, the length cannot be too large. When the length is reached $\overline{L}$, the ray is split (so the length is reduced to $\frac{\overline{L}}{2}$). The split ray (representing an angular range of width ϕ) continues to propagate until its length reaches $\overline{L}$.

The raycasting method can provide channel characteristics for the entire environment, i.e., many different receiver locations as well as a given transmitter location. In other words, once the location of the base station is determined, coverage, delay spread, and other channel characteristics can be calculated for the entire cell. The environment can be broken down into a number of several "slices" (regions of finite size, usually the same as the maximum effective area of the ray), and the interactions between all the slices can be calculated. Next, for each transmitter position, only the interaction between the transmitter and the "slice" that can be the initial interactor needs to be calculated.

3.2.1.1 Ray Tracing

The research on the ray tracing method was carried out very early, and this technology was first used to solve the prediction problem of long-distance radio wave propagation. Because this processing method needs to track every radio wave sent from the transmitter, people gave this deterministic modeling method this vivid name. Research on ray tracing was carried out very early. In 1959, Muldrew D proposed the ray-tracing method for the prediction of radio waves in the ionosphere, which was used to solve the problem of prediction of long-distance radio wave propagation [67], and he also used the term "Ray-Tracing" for the first time. Since then, some people have further studied the ray tracing technology of the ionosphere [68] and troposphere [69]. In 1968, Samson C et al. used ray tracing technology to calculate the diffraction loss when radio waves encounter obstacles [70]. Wang W Y D et al. studied the scattering problem with the ray tracing method [71]. References [72,73] propose methods to increase the speed of ray tracing calculations. In 1966, the time-domain finite difference algorithm was proposed [74], which provided a more effective numerical calculation tool for the ray tracing method. Ray tracing technology began to be widely used in the field of land mobile communications. It originated from the ray tracing program for indoor radio coverage prediction proposed by Motorola's wireless telephone system group in 1991. However, the program includes some unrealistic approximations, so it can only be carried out with quality predictions. The ray tracing model for the prediction of radio wave propagation in outdoor scenes was proposed and implemented in the early 1990s [75]. Reference [76] further uses multiple antennas for modeling and adds a discussion on receiver received power.

The simplest ray tracing model is the Two-Ray model [77,78], which can accurately reflect the propagation characteristics of the signal when there is only one direct path and one reflection path. Similar classic deterministic models such as the Ten-Ray model [79], Multi-Ray model [80], Advanced LOS path loss model [81], and Six-Ray model [82] can all be well realized for specific scenarios. Prediction: The establishment of this type of model does not start from the complex Maxwell equations, nor does it introduce FDTD calculations, but through sufficient observation of the scene, the main radio wave propagation mechanism (such as the number of paths, propagation mode, etc.) is stripped out, and the superposition principle is used to realize the prediction of signal strength. This method effectively reduces the computational complexity and the dependence on scene information, and can obtain better prediction model accuracy. However, this method has certain difficulties in realizing accurate and effective modeling in specific scenarios.

A method of approximately estimating the strength of high-frequency electromagnetic fields is based on the principle of geometric optics, by simulating the propagation path of rays (light) to determine reflection, refraction, etc. For the diffraction of obstacles, the geometrical optics theory is supplemented by introducing diffracted rays, that is, the theory of geometrical diffraction and uniform diffraction. The process of ray tracing is to scan all the surface elements of the emitted ray tube, determine the intersection point, then obtain the reflected ray, and then perform a new tracing process on the reflected ray until the ray exits the integral surface. Each ray has to go through this process. Therefore, as the size of the shape increases and the number of surface elements increases, the amount of calculation increases dramatically.

Traditional ray tracing methods determine all rays from a transmitter location to a receiver location. This method includes the following two steps [11].

In the first step, determine the energies of all rays from the transmitting position to the receiving position. This step is usually done by the mirror principle. Rays that arrive at the receiver by reflection generally behave the same as rays sent through a virtual mirror point (relative to the reflecting surface) of the original source (as shown in Figure 3.10).

In the second step, losses (caused by free space losses and finite reflection coefficients) are calculated, thus providing parameters for all multipath components.

Ray tracing can quickly calculate single and double reflections without the need for ray splitting. On the downside, the computational complexity grows exponentially as the number of reflections included in the simulation increases. Also, scattering and diffraction paths are not trivial. Finally, this method is not as efficient as the ray-emission method for calculating channel characteristics over a large range.

3.2.1.2 Effectiveness Considerations

Whether it is ray shooting or ray tracing, it is almost impossible to correctly predict the phase of arriving rays. Such a prediction would require geographic and building databases

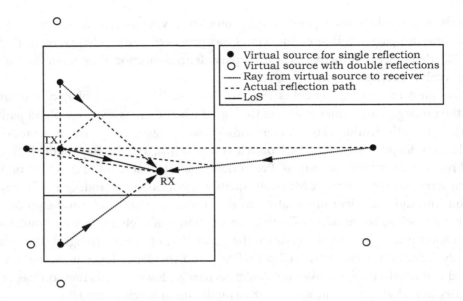

FIGURE 3.10   Mirroring principle [12].

within a fraction of a wavelength. Therefore, it is generally assumed that all rays have uniformly distributed random phases. In this case, only small-scale channel characteristics can be predicted; the impulse response is realized by adding random phases to the multipath components. This is yet another form of mixing determinism and randomness.

A way to further reduce computational complexity is to do ray tracing only in 2D dimensions, rather than 3D. The feasibility of this simplified method depends on the environment of the wave propagation [11].

- Indoor: Indoor environments usually need to consider three-dimensional dimensions in practical applications. Even when the base station and the mobile station are on the same floor, reflection paths through the ground and roof are still important propagation paths.

- Macrocell: By definition, the antenna of the base station needs to be significantly higher than the roof. The path typically passes over the roof to a point near the mobile station. Through these points, the rays travel further to the receiver, possibly by diffraction or by reflection from the opposite house wall. Under certain conditions, ray tracing on the vertical plane is sufficient, especially when ray tracing only needs to predict received power and delay spread. On the other hand, such a vertical ray tracing cannot correctly predict the direction of the rays reaching the receiver.

- Microcell, where the base station is far from the mobile station: In this case, the relative power of the rays propagating on the horizontal plane is small compared to the component above the roof. The horizontal component has experienced multiple diffraction and reflection processes, while the loss of the component above the roof is mostly determined by the diffraction loss around the base station and mobile station, so it has little relationship with the distance. In this case, a so-called 2.5-dimensional model can be used: on the one hand, only the components of the horizontal plane and on the other hand only the components of the vertical plane are simulated and the contributions of both are superimposed. 2.5D modeling can also be used for macrocells. However, whether in a macrocell or a microcell, some propagation paths cannot be accurately modeled by 2.5D ray tracing due to the height of the base station antenna close to the roof. For example, paths reflected from far-end interacting objects (high-rise buildings) are not included in this approach (as shown in Figure 3.11).

3.2.2 Geodatabase

The basis of all deterministic methods is geographic and morphological information. The accuracy of this information directly determines the accuracy that deterministic channel modeling can achieve.

For indoor environments, this information is usually available through building plans, which today is generally available in digital form.

In rural areas, the resolution of geographic databases is usually between 10 and 100 m. These databases are usually established in the form of satellite observations. Formal information (land use) is also available in many countries; however, obtaining this information in an automated and consistent manner is quite challenging.

In urban areas, digital databases use two different forms of data: vector data and pixel data: for vector data, the actual location of building endpoints is stored; for pixel data, a regular grid of points is added to the area. A pixel indicates whether it falls in "free space"

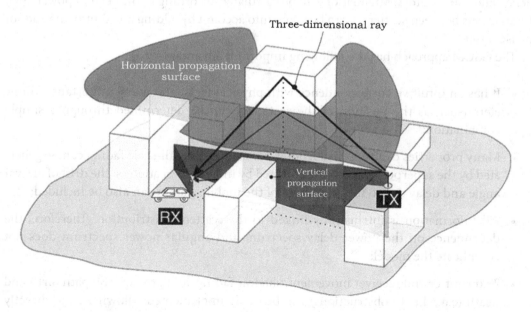

FIGURE 3.11   2D and 3D modeling diagram [12].

(street, park, etc.) or is covered by buildings. In both cases, building heights and materials can be included in the database.

3.3 GEOMETRY-BASED STOCHASTIC CHANNEL MODELING

In recent years, the random channel model based on geometry has been widely used. On the one hand, it has higher accuracy than the statistical model and can be better combined with MIMO technology; on the other hand, it has lower accuracy than the deterministic model computational complexity. Any geometry-based model is determined by the location of the scatterers. In deterministic geometric methods (such as ray tracing), the positions of scatterers are set in a database. In contrast, the Geometry-Based Stochastic Channel Model (GSCM) randomly generates the position of the scatterer according to some specific probability distribution, while the actual channel impulse response is realized by a simplified ray tracing method. The stochastic channel model based on geometry is introduced below.

3.3.1 First Order Scattering

GSCM was originally designed for channel simulation of multi-antenna systems on the base station side (diversity antennas and smart antennas). The original GSCM placed scatterers deterministically on a circle around the mobile station and assumed only first-order scattering (i.e., only one interacting body between the transmitter and receiver). About 20 years later, many literatures proposed that the first-order scattering model can be extended by arbitrarily placing scatterers. This arbitrary placement method can more effectively reflect physical facts. The assumption of first-order scattering also makes the process of ray tracing very simple: except for the direct path, all other paths are composed of two sub-paths connecting the scatterer and the transmitter and receiver respectively. These subpaths can characterize the angle of arrival, angle of departure, and propagation delay (and thus the total attenuation can be calculated according to the general power law). Interactions between scatterers can be taken into account by adding an additional random phase offset.

The GSCM approach has the following important advantages [83]:

- It has an intuitive correspondence with physical facts, and many important parameters (such as the position of the scatterer) can be determined through a simple combination;

- Many properties can be mined from it, for example, small-scale fading can be generated by the superposition of waves formed by individual scatterers; the drift of arrival angle and delay due to the movement of the mobile station can also be included;

- All information is intrinsically linked to the scatterer distribution. Therefore, the dependence on the power delay spectrum and angular power spectrum does not complicate the model;

- Transmitter and receiver movement, shadow fading, and propagation path birth and death (e.g., due to obstructions) can be easily implemented, allowing us to directly include long-term channel correlation properties.

The main difference between the different GSCM versions is the proposed scatterer distribution. The simplest GSCM is established by assuming that the scatterers are uniformly distributed in space. The energy contribution from distant scatterers is lower because these paths travel longer distances, resulting in more severe attenuation; this model is often referred to as the single-hop geometry model. Another way of thinking suggests randomly setting the scatterers around the mobile station. In the existing literature, other scatterer distributions around the mobile station have been further applied and analyzed; a unilateral Gaussian distribution related to the distance of the mobile station has obtained a more reasonable logarithmic attenuation PDP. To correlate scatterer density and intensity with distance, two implementations are possible. In the traditional method, the probability density equation of the scatterer is constantly adjusted, for example, the possibility of the scatterer appearing at a position farther away from the mobile station is less; while the "non-stationary scatterer intersection area" method is to combine the scatterer placed in the considered area according to a uniform distribution, the weight of their contributions is gradually reduced as the distance from the mobile station increases. For very high scatterer densities, the two methods are equivalent. However, the non-stationary scatterer cross-region method has many advantages, especially when the number of scatterers is limited, and the power delay spectrum has smaller statistical fluctuations.

Another important propagation characteristic comes from the presence of clusters of distant scatterers (e.g., large buildings, mountains, etc.). The farther scatterers will increase the time dispersion and angle dispersion, which will significantly affect the performance of MIMO systems. In GSCM, these can be taken into account by setting distant scatterer clusters at random positions in the cell [84], and the specific structure is shown in Figure 3.12.

3.3.2 Multi-Order Scattering

All of the above considerations are based on the assumption of only first-order scattering. This obviously limits the development of the model, since the position of the scatterer completely determines the angle of departure, angle of arrival, and delay, i.e., only two or

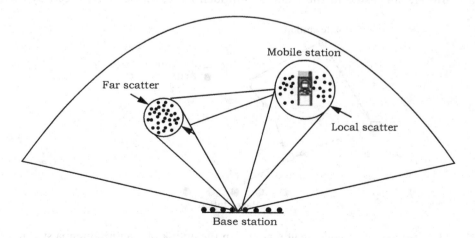

FIGURE 3.12   Structural principles of GSCM covering clusters of distal scatterers.

more of these parameters can be chosen independently. However, many scenarios (such as microcells or picocells) have multi-hop scattering where the angle of departure, angle of arrival, and delay are completely uncorrelated. In a microcell, the height of the base station is lower than the roof, so the wave propagation mainly passes through the waveguide in the street canyon, which contains many multiple reflections and diffractions (this phenomenon is more significant in the macrocell). For picocells, if the transmitter and receiver are in different rooms, the wave propagation can leave the transmitter's room by passing through walls, or through windows or doors, propagate through corridor waveguides, and diffract into the receiver's room.

If the directional channel characteristics only need to be reconstructed at one side of the communication terminal (for example, only using multiple antennas at the transmitting end or receiving end), multi-order scattering can be added to GSCM through the concept of equivalent scatterers. The position and path loss of these virtual first-order scatterers are realized by simulating the time delay and angle of arrival contributions of multi-order scattering (as shown in Figure 3.13). This approach is always possible because the time delay, horizon angle, and elevation angle of a first-order scatterer are always uniquely determined in a Cartesian coordinate system. Similar relationships also exist in statistical properties such as the joint angle-delay power spectrum and the probability density function of scatterer coordinates (such as spatial scatterer distribution) [85].

In MIMO systems, the concept of equivalent scatterers is invalid because the channel angular characteristics can only be correctly reconstructed in the case of one link. As a remedy, some literature proposes the use of double scattering, where the coupling between the scatterers around the base station and the scatterers around the mobile station is established by a so-called heuristic function (basically the spectrum of departure angles concerning this scatterer) stand up. It should be noted that the channel model in this method also considers some simple mechanisms to take waveguiding and diffraction into account.

Another way to introduce multi-order scattering into GSCM is to use the concept of paired scatterers suggested in the COST273 model. Here, the scatterers seen by the base

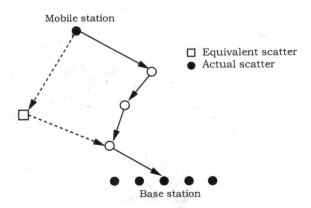

FIGURE 3.13 Example of an equivalent scatterer in the uplink of a multi-antenna system on the base station side.

station and the mobile station are different, and the coupling relationship of the double-ended scatterers is established through a random propagation link. This concept can also complete the decoupling of statistical parameters such as angle of arrival, angle of departure, and delay.

In time-varying channels (such as V2V channels), geometric models can be divided into two categories according to the distribution of scatterers: regular geometric models and irregular geometric models. Since the geometric model is more focused on the research of modeling methods, the modeling process does not have a clear scene distinction. Therefore, the following is an introduction from the perspective of modeling methods.

3.3.2.1 Regular Geometry Model

According to the distribution of scatterers, it is divided into different geometric structures such as single circle, double circle, ellipse, double cylinder, etc., and the following is briefly summarized according to the different geometric structures of scatterers. Reference [86] established a SISO dual-circle channel model including a direct path and a double-hop path, and established a V2V channel simulation model by using the sum-of-sinusoids method, and deduced the autocorrelation function, level Pass rate, and average fade duration. Reference [87] established a MIMO dual-circle narrowband channel model with isotropic and anisotropic scatterers with only double-hop paths in urban and suburban scenarios, and deduced the spatial Time correlation function, and based on this theoretical model, a statistical and deterministic simulation model is given. Reference [88] established a 3D MIMO dual-cylindrical channel model with anisotropic scatterers with only double-hop paths in urban scenarios. Based on this model, the impulse response function of the channel and the universal space-time correlation the closed solution of the function. Reference [89] established a 3D MIMO double-layer cylindrical Ricean fading wideband channel with anisotropic scatterers including direct path, single-hop (transmitter, receiver), and double-hop paths in urban and interstate highway scenarios. Based on the structure of this geometric model, the literature deduces the impulse response function, in which the distribution of the angle of arrival and the angle of departure adopts the von Mises distribution, and derives the space-time-frequency correlation function in the environment of 3D anisotropic scatterers, and the spatial Doppler Le power spectrum and power delay distribution, and the closed solution of the above function is obtained under the assumption that each multipath is independent. Reference [90] presents a 2D "double circle + ellipse" environment with anisotropic scatterers including direct paths, single-hop (transmitter, receiver) and double-hop paths in macrocell, microcell, and picocell scenarios. The MIMO V2V Ricean fading narrowband channel model is developed, by modifying different parameters in the model, and the model is suitable for different scenarios such as macrocells, microcells, and picocells. The established model considers the influence of vehicle density on channel characteristics. Based on this model, the literature presents a universal method to deduce the relationship between angle of arrival and angle of departure under different scatterer models such as single circle, double circle, and ellipse, and deduced the expressions of the space-time-frequency correlation function and spatial Doppler spectral density function Mode.

The advantage of the regular geometric model is that it does not need to rely on a large number of channel measurement results, but simulates the V2V channel based on the geometric structure assumptions and multipath statistical distribution in the propagation scenario, and establishes a theoretical analytical channel model. At the same time, based on the geometric model, the analytical analysis of the second-order statistical characteristics of the channel and the development of the channel simulation model can be conveniently carried out. However, the deficiency of the regular geometric model is simply assuming that the scatterers are distributed on a regular geometric structure, or that the statistics are distributed in a fixed interval. There is a certain gap between these simple assumptions and the actual environment, so this model is not convenient for carrying out traditional channel parameter characteristic analysis; at the same time, the complexity of such models is high. Figure 3.14 shows a schematic diagram of a typical double-cylindrical geometric model in regular geometric modeling. For specific parameter descriptions and model structure analysis, please refer to ref. [89].

3.3.2.2 Irregular Geometric Model
The basic idea of this method is to strive to build a more realistic model, in which the position and even the characteristics of the scatterers are defined, which is very close to the actual measurement results. First, the scatterers are divided into discrete scatterers

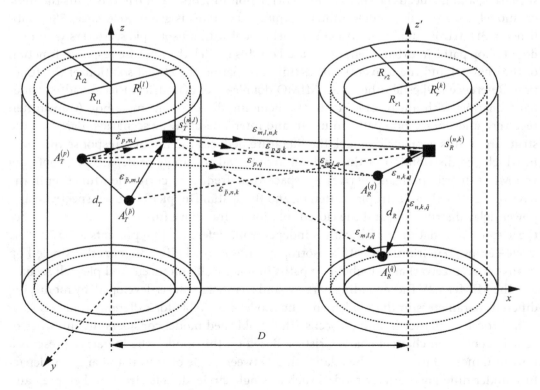

FIGURE 3.14 Schematic structure of the double-cylindrical geometric model [89].

and diffuse scatterers, where discrete scatterers include cars, houses, road signs, and other important scatter points distributed on both sides of the road, while diffuse scatterers are mainly some smaller distributions in the objects along the road, such as trees, etc. Finally, by using the ray tracing method, the total impulse response of the receiving end is obtained as the sum of individual impulse responses. Reference [91] presents this general modeling approach based on irregular geometry. The first is to sprinkle points. On the x-axis direction, the discrete scatterers are divided into moving and stationary scatterers. The distribution obeys the uniform distribution, and the distribution of the diffuse scatterers also obeys the uniform distribution. The probability densities of the three are different; in the y-axis direction, moving and stationary scatterers follow a Gaussian distribution, while diffuse scatterers follow a uniform distribution on both sides of the road. Then, according to the method of sprinkling points, the research of each parameter is carried out in the next step, and the impulse response of the final receiving end is given, which is the sum of the impulse responses of moving and stationary scatterers, LOS paths, and diffuse scatterers. Reference [92] used a general stochastic geometric modeling method to compare the autocorrelation function of the impulse response under different scattering methods, and set the speed and direction of the transmitter and receiver as random, and the scatterer. The speed and direction of the scatterers are also set to be random. Under this assumption, the impulse response of a reflection path is calculated and summed. Finally, the scatterers are divided into fast-moving and slow-moving scatterers. The fast-moving scatterers are set to three kinds of scatterer distributions; slow-moving scatterers are set to two kinds of scatterer distributions, and their impulse response autocorrelation functions are simulated and compared.

The irregular geometric model can effectively reflect the 3D characteristics of the channel, and it is easier to combine with the parameter analysis results of the measured data. It is also one of the main V2V modeling ideas at present. Figure 3.15 shows the schematic diagram of the irregular geometry sprinkle point model in a typical highway environment. For specific parameter definitions and model analysis, please refer to ref. [91].

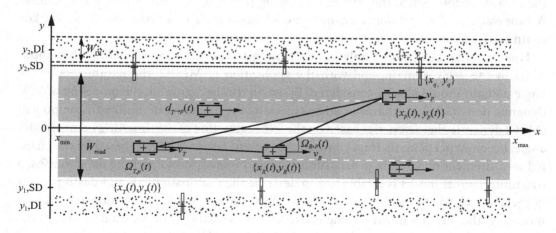

FIGURE 3.15   Irregular geometric sprinkler model [91].

3.4 TIME-VARYING NON-STATIONARY CHANNEL MODELING THEORY

Most of the parameters of the channel model mentioned above do not change with time, and cannot reflect the characteristics of time-varying non-stationary channels in high-speed mobile scenarios. In this section, V2V channels will be taken as an example to model the theory of time-varying non-stationary channels. Make an introduction.

When modeling the time-varying non-stationary characteristics of the channel, the common method is to use the Markov model to describe the multipath disappearance or increase caused by the sudden scatterer obstruction when the transmission environment changes unpredictably. The appearance of multipath due to scattered paths. Using a first-order two-state Markov chain to describe the continuation and jump of the multipath birth and death state can be expressed as:

$$TS = \begin{bmatrix} P_{00} & P_{01} \\ P_{10} & P_{11} \end{bmatrix} \quad SS = \begin{bmatrix} S_0 \\ S_1 \end{bmatrix} \tag{3.25}$$

Among them, TS is the state transition matrix, P_{ij} which represents the probability of i jumping from the current state to the state j. SS is a state-holding matrix, S_i which represents the existence probability of a state i in the entire state sequence, which satisfied $\sum_{i,j} P_{ij} = 1, \sum_{i} S_i = 1$. The research of this method is mainly to improve and innovate based on the classic TDL model. Reference [93] established a TDL channel model based on channel measurements of 3 V2V scenarios and 3 V2I scenarios of 5.9 GHz highway, canyon, and city, and established Rayleigh channel model and Ricean channel model under the assumption of generalized stationary non-correlation, and makes a detailed modeling of the complex Doppler effect of the vehicle channel. Reference [94] measured the channel characteristics of four large cities, one small city, and surrounding highways in Ohio at 5.2 GHz in detail. Based on this, a TDL channel model was established. According to the measurement results, the channel including power delay distribution was obtained. A time-varying non-stationary channel model based on a first-order two-state Markov chain can be obtained.

However, the TDL model only pays attention to the variation of multipath components in the delay domain, and with the application of MIMO, the parameters in the angle domain should also be considered. Based on this background, the modeling of V2V dynamic non-stationary characteristics, especially the modeling of multipath component (MPC) dynamic characteristics based on the joint extraction of the time-delay domain and angle, becomes particularly important. In the current research [95,96], based on the channel measurement of 5.3 GHz urban, suburban, and underground parking lot scenarios, a dynamic channel model is established to describe the multipath birth and death process of the channel, which includes the statistics of the following dynamic parameters. Sexual modeling: the number of MPCs, the existence time of MPCs, the location of MPCs, the energy of MPCs, and the fading status of MPCs. It should be pointed out that the dynamic channel modeling of time-varying non-stationary channels is still in a state of research at this stage, which includes extraction and clustering of time-varying multipath parameters, tracking of multipath clusters, and statistics of multipath clusters.

Research on V2V time-varying non-stationary channel characteristics has received extensive attention. Reference [97] analyzed the delay dispersion characteristics of V2V channels based on the channel characteristics of 5.2 GHz cities, small towns, and highway scenarios, and proposed a statistical non-stationary Channel impulse response model; however, the above work did not consider the distribution characteristics of the multipath angle space comprehensively. Reference [98] will consider only 5 MHz and 10 MHz bandwidth model extended to 1 MHz, 20 MHz, 33.3 MHz, and 50 TDL channel model at MHz bandwidth. Reference [99] analyzed the delay spread characteristics of V2V channels based on MIMO measurements in the 5.3 GHz frequency band, and found that the small-scale fading of non-stationary V2V channels obeys the Ricean distribution. Reference [100] studies V2V based on the measurement; the non-stationary characteristics of MIMO channels and the fading characteristics and cross-correlation characteristics of multipath components in the delay domain in the tapped delay line model of non-stationary channels are expounded. Reference [101] based on the V2V channel measurement in the 5.6 GHz frequency band selects special scenarios that are closely related to safety factors, including intersections, sheltered expressways, suburban merge areas, traffic congestion, tunnels and overpasses, etc., and analyzes different scenarios based on measurements. Statistical properties of the scene include local scattering function, power delay distribution, Doppler spread, root mean square delay spread, etc. Reference [102] proposed a bimodal Gaussian mixture model to describe the K-factor of the Ricean distribution that varies with time and frequency based on the channel measurement results of the above-mentioned special scenario in the 5.6 GHz band that is closely related to security factors.

To sum up, there are few studies on non-stationary channel models in current mobile scenarios. Traditional work still uses the idea of TDL model, and a small number of studies have carried out dynamic TDL modeling work. This is mainly limited by the lag in the development of dynamic channel modeling theory and the lack of directional channel tests. How to establish a more accurate multipath birth-death model in typical scenarios of related frequency bands needs to be further studied. Related work will help to propose a complete time-varying non-stationary 3D directional dynamic channel model to improve the simulation accuracy of time-varying channels. It truly reflects the characteristics of non-stationary channels.

3.5 MULTIPATH PARAMETER EXTRACTION METHOD

In recent years, with the wide application of MIMO technology in the field of wireless channel measurement and modeling and the continuous introduction of high-precision parameter estimation methods, the accuracy of wireless channel multipath parameter extraction has been continuously improved. The high-precision multipath parameter extraction algorithm helps to dig deep into the channel test data. The following will focus on channel multipath angle domain parameter estimation, analyze the basic data model, and introduce several common multipath parameter extraction methods [12].

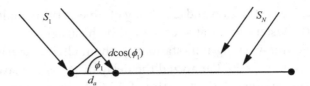

FIGURE 3.16   Schematic illustration of a plane wave incident from the angle $\phi_1,...,\phi_n$ in a uniform linear array [12].

Firstly, a mathematical model is established for the array and incoming signals. Taking a uniform linear array (ULA) containing N_r elements as an example for analysis, $r_n(t)$ represents the signal detected on the nth antenna. To simplify the discussion, it will be assumed that all waves propagate in a horizontal plane.

Assume that plane waves are incident on the array from N different directions (as shown in Figure 3.16), and each wave is described by its angle of arrival ϕ_i, and the relationship between the incident signal s_i and its received signal in the first antenna array element is:

$$r_1(t) = \sum_i a_{i,1} s_i(t - \tau_{i,1}) + n_1(t) \tag{3.26}$$

where $\tau_{i,1}$ is the propagation time from the ith source of the transmitted signal to the first antenna, $a_{i,1}$ is the (complex) amplitude of the signal, and $n_1(t)$ is the noise of the first antenna array element. Now consider the second antenna array element whose received signal is:

$$r_2(t) = \sum_i a_{i,2} s_i(t - \tau_{i,2}) + n_2(t) \tag{3.27}$$

If the i-th transmitted signal source is located in the far field, then there is $|a_{i,1}| = |a_{i,2}|$, and:

$$s_i(t - \tau_{i,2}) = s_i(t - \tau_{i,1}) \exp\left(-j(\tau_{i,2} - \tau_{i,1}) 2\pi f_c\right) \tag{3.28}$$

Equation (3.28) assumes that the signal is a narrowband signal on the RF sensing side, that is, the bandwidth of the signal is much smaller than the carrier frequency. The physical explanation for this phenomenon is that the only effect brought by the antenna position is the additional time delay caused by the phase offset. This delay difference can be expressed as:

$$\tau_{i,2} - \tau_{i,1} = \left(\frac{d_a}{c_0}\right) \cos(\phi_i) \tag{3.29}$$

The relationship between r_2 and s can be expressed as:

$$r_1(t) = \sum_i \tilde{s}_i(t) + n_1(t) \tag{3.30}$$

$$r_2(t) = \sum_i \tilde{s}_i(t) \exp(-j2\pi d_a \cos(\phi_i)/\lambda_0) + n_2(t) \tag{3.31}$$

Among them, $\tilde{s}_i(t) = a_{i,1} s_i(t - \tau_{i,1})$. For the next antenna array element, one obtains:

$$r_3(t) = \sum_i \tilde{s}_i(t) \exp(-j2\pi 2 d_a \cos(\phi_i)/\lambda_0) + n_3(t) \tag{3.32}$$

the general relationship between r and s can be summarized, namely:

$$r(t) = As(t) + n(t) \tag{3.33}$$

Where, $r(t) = [r_1(t), r_2(t), \cdots, r_{N_r}(t)]^T$, $s(t) = [s_1(t), s_2(t), \cdots, s_N(t)]^T$,

$n(t) = [n_1(t), n_2(t), \cdots, n_{N_r}(t)]^T$,

$$A = \begin{pmatrix} 1 & 1 & \cdots & 1 \\ \exp(-jk_0 d_a \cos(\varphi_1)) & \exp(-jk_0 d_a \cos(\varphi_2)) & \cdots & \exp(-jk_0 d_a \cos(\varphi_N)) \\ \exp(-j2k_0 d_a \cos(\varphi_1)) & \exp(-j2k_0 d_a \cos(\varphi_2)) & \cdots & \exp(-j2k_0 d_a \cos(\varphi_N)) \\ \vdots & \vdots & \ddots & \vdots \\ \exp(-j(N_r-1)k_0 d_a \cos(\varphi_1)) & \exp(-j(N_r-1)k_0 d_a \cos(\varphi_2)) & \cdots & \exp(-j(N_r-1)k_0 d_a \cos(\varphi_N)) \end{pmatrix} \tag{3.34}$$

Equation (3.34) is the steering matrix. Furthermore, it is assumed that the noise on different antenna array elements is independent, so the correlation matrix of the noise is a diagonal matrix with input parameters σ_n^2 on the main diagonal. Based on this definition, several common multipath parameter extraction methods will be introduced next.

3.5.1 Beamforming Method

The easiest way to determine the multipath angle of incidence is to perform Fourier transform on the signal vector r. This method can obtain the angle of arrival with an angular resolution determined by the size of the array, approximately $\dfrac{2\pi}{N_r}$. The advantage of this method is that the implementation process is simple (only fast Fourier transform is required); the disadvantage is that the resolution is low. Specifically, the calculation formula of the angle spectrum $P_{BF}(\phi)$ can be expressed as:

$$P_{BF}(\phi) = \frac{\alpha^\dagger(\phi) R_{rr} \alpha(\phi)}{\alpha^\dagger(\phi) \alpha(\phi)} \tag{3.35}$$

where R_{rr} is the correlation matrix of the incident signal, and:

$$\boldsymbol{\alpha}_{RX}(\phi) = \begin{pmatrix} 1 \\ \exp(-jk_0 d_a \cos(\phi)) \\ \exp(-j2k_0 d_a \cos(\phi)) \\ \vdots \\ \exp(-j(N_r-1)k_0 d_a \cos(\phi)) \end{pmatrix} \tag{3.36}$$

ϕ is the steering matrix in the direction.

3.5.2 High-Precision Algorithm

The problem of too low estimation resolution in channel multipath extraction can be solved by adopting a high-resolution algorithm. The resolution of these algorithms is not limited by the size of the antenna array, but only by modeling errors and noise. The price paid for the advantages of such algorithms is computational complexity. In addition, the high-precision algorithm usually has a certain limit on the number of multipath components in the direction to be estimated during the implementation process. Typical high-resolution algorithms mainly include the following [12].

- ESPRIT (Estimation of Signal Parameters by Rotational Invariance Techniques) algorithm: This algorithm determines the subspace of the signal and extracts the closed form of the direction of arrival. This algorithm is mainly used for linear antenna arrays.

- MUSIC (Multiple Signal Classification) algorithm: This algorithm needs to determine the signal and noise subspaces, but then uses a spectral search method to find the direction of the arrival angle [103].

- Capon's minimum variance method (Minimum Variation Method, MVM): This method is a pure spectrum search method, which determines an angle spectrum for the direction considered, so that the sum of noise and interference from other directions is minimized. Its improved angle spectrum is easy to calculate and can be expressed as:

$$P_{MVM}(\phi) = \frac{1}{\boldsymbol{\alpha}^+(\phi)\boldsymbol{R}_{rr}^{-1}\boldsymbol{\alpha}(\phi)} \tag{3.37}$$

- Maximum likelihood estimation of incident wave parameters: The problem with maximum likelihood parameter extraction is high computational complexity. An efficient iterative algorithm is the Space Alternating Generalized Expectation (SAGE) maximization algorithm [104]. Since 2000, this algorithm has become the most popular method for channel sounding estimation. Its disadvantage is that it may converge to the local optimal value during the iterative process, rather than the global optimal value.

In order to introduce multipath parameter extraction more vividly, the following uses the SAGE algorithm to carry out an example of multipath parameter extraction and channel reconstruction based on the channel measurement data of a large-scale antenna array in a large conference room scene under the condition of a center carrier frequency of 6 GHz [105]. Figure 3.17 shows the measured and estimated power delay spectrum results using the SAGE algorithm. It can be seen that the two have a good consistency. Figure 3.18 further shows the extraction results of the angle-of-off angle power spectrum. It can be seen that most of the main multipath signals can be obtained from it.

A problem common to all high high-precision algorithms is the calibration of the antenna array [106]. Most algorithms make assumptions about the array: the antenna patterns of all the elements are independent, and there is no mutual coupling between the

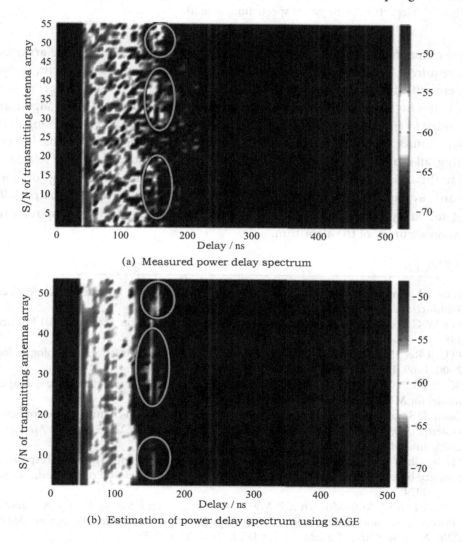

(a) Measured power delay spectrum

(b) Estimation of power delay spectrum using SAGE

FIGURE 3.17   Power delay spectrum schematic.

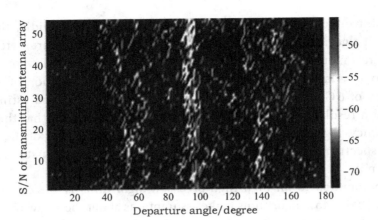

FIGURE 3.18   Departure angle power spectrum schematic.

antenna elements. In order to obtain reasonable estimation results, precise array calibration is required. This calibration must be repeated because temperature drift, component aging, etc. will affect the calibration [107].

For many high-resolution algorithms (including some subspace-based algorithms), it is necessary to ensure that the correlation matrix is non-singular [12]. The singularity R_{rr} is usually caused by the correlation of waves incident in different directions. For channel sounding, all signals typically come from the same source and are therefore related to each other. In this case, methods such as subarray averaging ("spatial smoothing" or "forward-backward" averaging) need to be used to obtain a reasonable correlation matrix [49]. The lack of sub-array averaging reduces the effective size of the array, thereby reducing the estimation accuracy of the algorithm.

REFERENCES

[1] Hata M. Empirical formulas for propagation loss in land mobile radio service. *IEEE Transactions on Vehicular Technology*, 1980, 29(3): 317–325.

[2] Lee W C Y. *Mobile Communications Engineering*. New York: McGraw Hill Publications, 1985.

[3] ITU-R Rec. M.1225, Guidelines for evaluation of radio transmission technologies for IMT-2000, 1–60, 1997.

[4] 3GPP TR25.996 V17.0.0, Technical specification group radio access networks; spatial channel model for MIMO simulations (Release 17), 1–38, 2022.

[5] Baum D S, Salo J, Galdo G D, Milojevic M, Kyösti P. An interim channel model for beyond-3G systems. *2005 IEEE 61stVehicular Technology Conference (VTC 2005-Spring), May 30–June 1, 2005, Stockholm, Sweden*. Piscataway, NJ: IEEE Press, 2005.

[6] He R, Chen W, Ai B, et al. On the clustering of radio channel impulse responses using sparsity-based methods. *IEEE Transactions on Antennas and Propagation*, 2016, 64(6): 2465–2474.

[7] He R, Chen W, Ai B, Molisch A F, Wang W, Zhong Z, Yu J, Sangodoyin S. A sparsity-based clustering framework for radio channel impulse responses. *IEEE VTC-Spring, May 15–18, 2016, Nanjing, China*. Piscataway, NJ: IEEE Press, 2016: 1–5.

[8] He R, Li Q, Ai B, Li-Ao Geng Y, Molisch A F, Kristem V, Zhong Z, Yu J. An automatic clustering algorithm for multipath components based on Kernel-power-density. *IEEE WCNC, March 19-22, 2017, San Francisco, USA.* Piscataway, NJ: IEEE Press, 2017: 1–6.

[9] Erceg V, IEEE 802.11-03/940r4 TGn channel models, 1–50, 2004.

[10] WINNER II D1.1.2 V1.0, WINNER II channel models, 1–82, 2007.

[11] Saunders S R, Zavla A A. *Antennas and Propagation for Wireless Communication Systems.* 2nd ed. Chichester: Wiley, 2007.

[12] Molisch A F. *Wireless Communication.* 2nd ed. Hoboken, NJ: Wiley, 2010.

[13] Rappaport T S. *Wireless Communications: Principles and Practice.* 2nd ed. Upper Saddle River, NJ: Prentice Hall PTR, 2001.

[14] Okumura Y, Ohmori E, Kawano T, et al. Field strength and its variability in VHF and UHF land-mobile radio service. *Review of the Electrical Communication Laboratory,* 1968, 16(9–10): 825–873.

[15] Hata M. Empirical formula for propagation loss in land mobile radio services. *IEEE Transactions on Vehicular Technology,* 1980, 29(3): 317–325.

[16] Walfisch J, Bertoni H L. A theoretical model of UHF propagation in urban environments. *IEEE Transactions on Antennas and Propagation,* 1988, 36(12): 1788–1796.

[17] Feuerstein M J, Blackard K L, Rappaport T S, et al. Path loss delay spread, and outage models as functions of antenna height for microcellular system design. *IEEE Transactions on Vehicular Technology,* 1994, 43(3): 487–498.

[18] Devasirvatham D M J, Banerjee C, Krain M J, Rappaport D A. Multi-frequency radiowave propagation measurements in the portable radio environment. *IEEE International Conference on Communications, April 16–19, 1990, Atlanta, GA, USA.* Piscataway, NJ: IEEE Press, 1990: 1334–1340.

[19] Rustako A J, Amitay N, Oweans G J, et al. Radio propagation at microwave frequencies for line-of-sight microcellular mobile and personal communications. *IEEE Transactions on Vehicular Technology,* 1991, 40(1): 203–210.

[20] Lee W Y C. *Mobile Cellular Telecommunications Systems.* New York: McGraw-Hill, 1989.

[21] Edwards R, Durkin J. Computer prediction of service areas for VHF mobile radio networks. *Proceeding of IEEE,* 1969, 116(9): 1493–1500.

[22] Goldsmith A J, Greenstein L J. A measurementbased model for predicting coverage areas of urban microcells. *IEEE Journal on Selected Areas in Communications,* 1993, 11(7): 1013–1023.

[23] Berg J-E. A recursive method for street microcell path loss calculations. *Sixth IEEE International Symposium on Personal, Indoor and Mobile Radio Communications (PIMRC'95), Wireless: Merging onto the Information Superhighway, September 27–29, 1995, Toronto, Canada.* Piscataway, NJ: IEEE Press, 1995: 140–143.

[24] Lee W C Y, Lee D J Y. Microcell prediction in dense urban area. *IEEE Transactions on Vehicular Technology,* 1998, 47(1): 246–253.

[25] Kurner T, Neuland M. Application of Bertoni's work to propagation models used for the planning of real 2G and 3G cellular networks. *3rd European Conference on Antennas and Propagation, March 23–27, 2009, Berlin, Germany.* Piscataway, NJ: IEEE Press, 2009: 1686–1690.

[26] Alexander S E. Radio propagation within buildings. *Electronics Letters,* 1982, 18(21): 913–914.

[27] Rappapon T S. Characterization of UHF multipath radio channels in factory buildings. *IEEE Transactions on Antennas and Propagation,* 1989, 37(8): 1058–1069.

[28] Cox D C, Murray R, Norris A. Measurements of 800 MHz radio transmission into buildings with metallic walls. *The Bell System Technical Journal,* 1983, 62(9): 2695–2717.

[29] Devasirvatham D J, Krain M J, Rappaport D A. Radio propagation measurements at 850 MHz and 4.0 GHz inside two dissimilar office buildings. *Electronics Letters,* 1990, 26(7): 445–447.

[30] Rappaport T S, Seidel S Y, Takamizawa K. Statistical channel impulse response models for factory and open plan building radio communicate system design. *IEEE Transactions on Wireless Communications*, 1991, 39(5): 794–807.

[31] Seidel S Y, Rappaport T S. 914 MHz path loss prediction models for indoor wireless communications in multifloored buildings. *IEEE Transactions on Antennas and Propagation*, 1992, 40(2): 207–217.

[32] Saleh A A, Valenzuela R. A statistical model for indoor multipath propagation. *IEEE Journal on Selected Areas in Communications*, 1987, 5(2): 128–137.

[33] Steinbauer M, Molisch A F, Bonek E. The double-directional radio channel. *IEEE Antennas and Propagation Magazine*, 2001, 43(4): 51–63.

[34] Fuhl J, Molisch A F, Bonek E. Unified channel model for mobile radio systems with smart antennas. *IEEE Proceedings-Radar, Sonar and Navigation*, 1998, 145(1): 32–41.

[35] Laurila J, Molisch A F, Bonek E. Influence of the scatter distribution on power delay profiles and azimuthal power spectra of mobile radio channels. *5th International Symposium on Spread Spectrum Techniques and Applications, September 4, 1998, Sun City, South Africa.* Piscataway, NJ: IEEE Press, 1998: 267–271.

[36] Kermoal J P, Schumacher L, Pedersen K I, et al. A stochastic MIMO radio channel model with experimental validation. *IEEE Journal on Selected Areas in Communications*, 2002, 20(6): 1211–1226.

[37] Rice P L. Transmission loss predictions for tropospheric communication circuits[M]. US Government Printing Office, 1965.

[38] Longley A G. Prediction of tropospheric radio transmission loss over irregular terrain: A computer method-1968[M]. Institute for Telecommunication Sciences, 1968.

[39] Longley A G. Radio propagation in urban areas[C]//28th ieee vehicular technology conference. IEEE, 1978, 28: 503–511.

[40] Weiner M M. Use of Longley-Rice and Johnson-Gierhart tropospheric radio propagation programs: 0.02–20 GHz. *IEEE Journal on Selected Areas in Communications*, 1986, 4(2): 297–307.

[41] Dadson C E, Durkin J, Martin E. Computer prediction of field strength in the planning of radio systems. *IEEE Transactions on Vehicular Technology*, 1975, 24(1): 1–7.

[42] Weissberger M, Meidenbaue R R, Riggins H, et al. *Radio Wave Propagation: A Handbook of Practical Techniques for Computing Basic Transmission Loss and Field Strength: DTIC Accession Number AD A122090.* Annapolis, MD: IIT Research Institute Report to Electronic Compatibility Analysis Center, 1982.

[43] Sciandra R M. *TIREM/SEM Programmer's Reference Manual: ECAC-CR-90-039.* Annapolis, MD: Electromagnetic Compatibility Analysis Center Report, 1990.

[44] Ayasli S. SEKE: A computer model for low altitude radar propagation over irregular terrain. *IEEE Transactions on Antennas and Propagation*, 1986, 34(8): 1013–1022.

[45] Ayasli S, Carlson M B. *SEKE: A Computer Model For Low-Altitude Radar Propagation over Irregular Terrain: CMT-70.* Cambridge, MA: MIT/Lincoln Lab, 1985.

[46] Gierhart G D, Johnson M E. *Computer Programs for Air/Ground Propagation and Interference Analysis, 0.1 to 20 GHz: AD 770335.* Springfield, VA: National Technical Information Service, 1973.

[47] Johnson M E, Gierhart G D. *Applications Guide for Propagation and Interference Analysis Computer Programs (0.1 to 20 GHz)[R].* Department of Transportation, Federal Aviation Administration, Systems Research and Development Service. Washington, DC: Federal Aviation Administration, 1978.

[48] Gierhart G D, Johnson M E. Propagation Model (O. 1 to 20 GHz) Extensions for 1977 Computer Programs[R]. Department of Transportation, Federal Aviation Administration, Systems Research & Development Service. Washington, DC: Federal Aviation Administration, 1978.

[49] Balth van der Pol, Bremmer H. The propagation of radio waves over a finitely conducting earth. *Philosophical Magazine*, 1938(25): 817–934.

[50] Ji Z, Li B H, Wang H X, et al. Efficient ray-tracing methods for propagation prediction for indoor wireless communications. *IEEE Transactions on Antennas and Propagation*, 2001, 43(2): 41–49.

[51] O'Brien W M, Kenny E M, Cullen P J. An efficient implementation of a three-dimensional microcell propagation tool for indoor and outdoor urban environments. *IEEE Transactions on Vehicular Technology*, 2000, 49(2): 622–630.

[52] Passerini C. A quality measure for ray-tracing algorithms. *IEEE Transactions on Antennas and Propagation*, 2001, 49(3): 500–502.

[53] Dottling M, Jahn A, Didascalou D, et al. Two-and three-dimensional ray tracing applied to the land mobile satellite (LMS) propagation channel. *IEEE Transactions on Antennas and Propagation*, 2001, 43(6): 27–36.

[54] Ott R H, Berry L A. An alterative integral equation for propagation over irregular terrain. *Radio Science*, 1970, 5(5): 767–771.

[55] Ott R H. An alterative integral equation for propagation over irregular terrain, 2. *Radio Science*, 1971, 6(4): 429–435.

[56] Janaswamy R. A fredholm integral equation method for propagation predictions over small terrain irregularities. *IEEE Transactions on Antennas and Propagation*, 1992, 40(11): 1416–1422.

[57] Akorli F K, Costa E. An efficient solution of an integral equation applicable to simulation of propagation along irregular terrain. *IEEE Transactions on Antennas and Propagation*, 2001, 49(7): 1033–1036.

[58] Brennan C, Cullen P J, Rossi L. An MFIE-based tabulated interaction method for UHF terrain propagation problems. *IEEE Transactions on Antennas and Propagation*, 2000, 48(6): 1003–1005.

[59] Schuster J W, Luebbers R J. Comparison of GTD and FDTD predictions for UHF radio wave propagation in a simple outdoor urban environment. *IEEE Antennas and Propagation Society International Symposium, July 13–18, 1997, Montreal, Quebec, Canada*. Piscataway, NJ: IEEE Press, 1997: 2022–2025.

[60] Hong W, Liu Y W, Mei K K. Application of the measured equation of invariance to solve scattering problems involving penetrable medium. *Radio Science*, 1994, 29(4): 897–906.

[61] Jevtic J, Lee R. A theoretical and numerical analysis of the measured equation of invariance. *IEEE Transactions on Antennas and Propagation*, 1994, 42(8): 1097–1105.

[62] Chen J, Hong W. Iterative procedure in matrix form based on MEI for scattering by multi-cylinders. *Electronics Letters*, 1996, 32(12): 1072–1074.

[63] Marcus S W. A model to calculate EM fields in tropospheric duct environments at frequencies through SHF[J]. *Radio Science*, 1982, 17(5): 895–901.

[64] Blaunstein N, Giladi R, Levin M. Characteristic's prediction in urban and suburban environments. *IEEE Transactions on Vehicular Technology*, 1998, 47(1): 225–234.

[65] Mogensen P, Wigard J. *Digital Mobile Radio – The View of COST 231*. Luxembourg: European Union, 1999.

[66] Muldrew D. An ionospheric ray-tracing technique and its application to a problem in a long-distance radio propagation. *IRE Transactions on Antennas and Propagation*, 1959, 7(4): 393–396.

[67] Shmoys J. Ray tracing in the ionosphere. *Electronics Letters*, 1968, 4(15): 302–304.

[68] Vogler L. An approximate height formula for tropospheric ray tracing. *IEEE Transactions on Antennas and Propagation*, 1971, 19(6): 794–796.

[69] Samson C, Peterson C, Farrow J, et al. A ray-tracing analysis of a prolonged fadeout on an obstacle-diffraction radio path. *IEEE Transactions on Communication Technology*, 1968, 16(3): 445–450.

[70] Wang W Y D, Deschamps G A. Application of complex ray tracing to scattering problems. *Proceedings of the IEEE*, 1974, 62(11): 1541–1551.

[71] Fujimoto A, Tanaka T, Iwata K. ARTS: Accelerated ray-tracing system. *IEEE Computer Graphics and Applications*, 1986, 6(4): 16–26.

[72] Plunkett D J, Bailey M J. The vectorization of a ray-tracing algorithm for improved execution speed. *IEEE Computer Graphics and Applications*, 1985, 5(8): 52–60.

[73] Yee K S. Numerical solution of initial boundary value problems involving Maxwell's equations in isotropic media. *IEEE Transactions on Antennas and Propagation*, 1966, 14(3): 302–307.

[74] Lebherz M, Wiesbeck W, Krank W. A versatile wave propagation model for the VHF/UHF range considering three-dimensional terrain. *IEEE Transactions on Antennas and Propagation*, 1992, 40(10): 1121–1131.

[75] Knörzer S, Baldauf M A, Fugen T, Wiesbeck W. Channel analysis for an OFDM-MISO train communications system using different antennas. *2007 IEEE 66th Vehicular Technology Conference, September 30–October 3, 2007, Baltimore, MD, USA*. Piscataway, NJ: IEEE Press, 2007: 809–813.

[76] Causebrook J H. Vodafone's fieldstrength prediction method: COST 231 TD (93)1. Barcelona, Spain, 1993: 1–5.

[77] Wagen J F. SIP simulation of UHF propagation in urban microcells. *The 41st IEEE Vehicular Technology Conference, May 19–22, 1991, St. Louis, USA*. Piscataway, NJ: IEEE Press, 1991.

[78] Amitay N. Modeling and computer simulation of wave propagation in linear line-of-sight microcells. *IEEE Transactions on Vehicular Technology*, 1992, 41(4): 337–342.

[79] Tam W K, Tran V N. Multi-ray propagation model for indoor wireless communications. *Electronics Letters*, 1996, 32(2): 135–137.

[80] Oda Y, Tsunekawa K, Hata M. Advanced LOS path-loss model in microcellular mobile communications. *IEEE Transactions on Vehicular Technology*, 2000, 49(6): 2121–2125.

[81] Wang Z, Jin R, Jin Y. Path loss prediction for mobile digital TV propagation under viaduct. *IEEE Transactions on Broadcasting*, 2011, 57(1): 37–45.

[82] Molisch A F, Kuchar A, Laurila J, et al. Geometry-based directional model for mobile radio channels-principles and implementation. *European Transactions on Telecommunications*, 2003, 14(4): 351–359.

[83] Almers P, Bonek E, Burr A, et al. Survey of channel and radio propagation models for wireless MIMO systems. *EURASIP Journal on Wireless Communications and Networking*, 2007, 2007(1): 56–75.

[84] Wang L C, Liu W C, Cheng Y H. Statistical analysis of a mobile-to-mobile rician fading channel model. *IEEE Transactions on Vehicular Technology*, 2009, 58(1): 32–38.

[85] Patzold M, Hogstad B O, Youssef N. Modeling, analysis, and simulation of MIMO mobile-to-mobile fading channels. *IEEE Transactions on Wireless Communications*, 2008, 7(2): 510–520.

[86] Zajic A G, Stuber G L. A three-dimensional MIMO mobile-to-mobile channel model. *IEEE Wireless Communications and Networking Conference, March 11-15, 2007, Kowloon, Hong Kong, China*. Piscataway, NJ: IEEE Press, 2007: 1883–1887.

[87] Zajic A G, Stuber G L. Three-dimensional modeling and simulation of wideband MIMO mobile-to-mobile channels. *IEEE Transactions on Wireless Communications*, 2009, 8(3): 1260–1275.

[88] Cheng X, Wang C X, Laurenson D I, et al. An adaptive geometry-based stochastic model for non-isotropic MIMO mobile-to-mobile channels. *IEEE Transactions on Wireless Communications*, 2009, 8(9): 4824–4835.

[89] Karedal J, Tufvesson F, Czink N, et al. A geometry-based stochastic MIMO model for vehicle-to-vehicle communications. *IEEE Transactions on Wireless Communications*, 2009, 8(7): 3646–3657.

[90] Borhani A, Patzold M. Modeling of vehicle-to-vehicle channels in the presence of moving scatterers. *IEEE Vehicular Technology Conference (VTC Fall), September 3–6, 2012, Quebec City, Canada*. Piscataway, NJ: IEEE Press, 2012: 1–5.

[91] Acosta-Marum G, Ingram M A. Six time-and frequency-selective empirical channel models for vehicular wireless LANs. *IEEE Vehicular Technology Magazine*, 2007, 2(4):4–11.

[92] Matolak D W, Sen I, Xiong W. Channel modeling for V2V communications. *2006 Third Annual International Conference on Mobile and Ubiquitous Systems: Networking & Services, July 17–21, 2006, San Jose, CA, USA.* Piscataway, NJ: IEEE Press, 2006: 1–7.

[93] He R, Renaudin O, Kolmonen V M, et al. A dynamic wideband directional channel model for vehicle-to-vehicle communications. *IEEE Transactions on Industrial Electronics*, 2015, 62(12): 7870–7882.

[94] He R, Renaudin O, Kolmonen V M, Haneda K, Zhong Z, Ai B, Oestges C. Statistical characterization of dynamic multi-path components for vehicle-to-vehicle radio channels. *IEEE VTC-Spring, May 11–14, 2015, Glasgow, United Kingdom.* Piscataway, NJ: IEEE Press, 2015: 1–6.

[95] Sen I, Matolak D W. Vehicle-vehicle channel models for the 5-GHz band. *IEEE Transactions on Intelligent Transportation Systems*, 2008, 9(2): 235–245.

[96] Qiong W, Matolak D W, Sen I. 5-GHz-band vehicle-to-vehicle channels: models for multiple values of channel bandwidth. *IEEE Transactions on Vehicular Technology*, 2010, 59(5): 2620–2625.

[97] Renaudin O, Kolmonen V M, Vainikainen P, Oestges C. Wideband MIMO car-to-car radio channel measurements at 5.3 GHz. *The 68th IEEE Vehicular Technology Conference (VTC 2008-Fall), September 21–24, 2008, Calgary, BC, Canada.* Piscataway, NJ: IEEE Press, 2008: 1–5.

[98] Renaudin O, Kolmonen V M, Vainikainen P, Oestges C. Car-to-car channel models based on wideband MIMO measurements at 5.3 GHz. *3rd European Conference on Antennas and Propagation, March 23–27, 2009, Berlin, Germany.* Piscataway, NJ: IEEE Press, 2009: 635–639.

[99] Bernadó L, Zemen T, Tufvesson F, et al. Delay and Doppler spreads of nonstationary vehicular channels for safety-relevant scenarios. *IEEE Transactions on Vehicular Technology*, 2014, 63(1): 82–93.

[100] Bernadó L, Zemen T, Tufvesson F, et al. Time-and frequency-varying K-factor of non-stationary vehicular channels for safety-relevant scenarios. *IEEE Transactions on Intelligent Transportation Systems*, 2015, 16(2): 1007–1017.

[101] Schmidt R O. Multiple emitter location and signal parameter estimation. *IEEE Transactions on Antennas and Propagation*, 1986, 34(3): 276–280.

[102] Fleury B H, Tschudin M, Heddergott R, et al. Channel parameter estimation in mobile radio environments using the SAGE algorithm. *IEEE Journal on Selected Areas in Communications*, 1999, 17(3): 434–450.

[103] Wang Q, Ai B, He R, Li J, Zhong Z, Li N, Qin H. A research on SAGE algorithm based on massive MIMO channel measurements[C]//2017 *XXXIInd General Assembly and Scientific Symposium of the International Union of Radio Science (URSI GASS)*. IEEE, 2017: 1–4.

[104] Wang T, Ai B, He R, et al. Two-dimension direction-of-arrival estimation for massive MIMO systems. *IEEE Access*, 2015, 3(11): 2122–2128.

[105] He R, Molisch A F, Tufvesson F, et al. Vehicle-to-vehicle propagation models with large vehicle obstructions. *IEEE Transactions on Intelligent Transportation Systems*, 2014, 15(5): 2237–2248.

[106] Haardt M, Nossek J A. Unitary ESPRIT: how to obtain increased estimation accuracy with a reduced computational burden. *IEEE Transactions on Signal Processing*, 1995, 43(5): 1232–1242.

[107] He R, Zhong Z, Ai B, et al. Simplified analytical propagation model for railway environments based on uniform theory of diffraction. *Electronics Letters*, 2013, 49(6): 397–399.

Highway Traffic Scenario Channel Measurement and Modeling

4.1 OVERVIEW

With the development of Intelligent Transportation Systems (ITS), research on V2V wireless communication systems in highway traffic environments has gained increasing attention in recent years. Future ITS aims to achieve real-time interaction and global sharing of safety information among vehicles on the road through wireless transmission using Vehicular Ad Hoc Networks (VANETs), which will effectively ensure driving safety. Building upon the IEEE 802.11p standard, industry standards for wireless access technologies in V2V environments have been proposed in the United States and Europe. In 2014, the National Highway Traffic Safety Administration (NHTSA) in the United States issued a statement, anticipating that by 2020, US vehicles would be mandated to install V2V wireless communication modules to enhance driving safety.

The design of V2V wireless communication systems relies on a deep understanding of the V2V wireless propagation channel. The V2V environment is complex and dynamic, with a significant presence of dynamic scattering objects around the communication link. Additionally, due to the low height of vehicle-mounted antennas (approximately 2 m), the movement of large vehicles (such as trucks and buses) can frequently block the line-of-sight (LOS) link, thereby degrading signal transmission quality. Therefore, the radio wave propagation characteristics in V2V environments differ from those of traditional cellular network macrocell channels. There is an urgent need for in-depth research on V2V wireless propagation channels to serve the design of V2V wireless communication systems effectively.

Most of the research on V2V wireless channels focuses on traditional driving environments, such as highways, urban, suburban, and rural areas. Due to testing difficulties, researchers have mainly focused on LOS scenarios in their studies, overlooking the

DOI: 10.1201/9781032669793-4

occasional occurrence of non-LOS (NLOS) propagation during the driving process. While it reduces model complexity and statistically reflects the characteristics of V2V wireless channels, the measurement results cannot scientifically reflect the impact of NLOS scenarios on path loss and delay spread. They also cannot be used to reasonably assess the quality of V2V communication links when NLOS propagation occurs, compromising the reliability of wireless signal transmission during driving hazards (such as occlusion of the driver's line of sight due to the presence of other vehicles or buildings, leading to NLOS propagation scenarios). Therefore, it is necessary to conduct targeted channel measurement and modeling specifically for complex NLOS scenarios in the V2V environment. In these scenarios, NLOS propagation occurs primarily due to two reasons.

- Obstructions by large vehicles: when large vehicles such as trucks or buses enter the communication path between cars, they block the LOS link, leading to NLOS conditions.

- Separation of driving routes at street crossings: when vehicles encounter crossroads and their driving paths diverge, the LOS link can be obstructed by roadside obstacles, causing NLOS conditions.

In these cases, as the LOS propagation link is spatially disrupted, the signal experiences more severe shadowing effects and time dispersion, leading to changes in wireless channel characteristics. Studying the wireless channel propagation characteristics in the scenarios mentioned above can help improve the reliability of the V2V wireless communication system design.

In the traditional sense, most wireless channels comply with the Wide Sense Stationary (WSS) assumption [2], which allows for a significant simplification of channel models [3]. However, it no longer holds due to the dynamic and time-varying nature of the scattering environment in V2V scenarios [4]. In references [5–7], analytical channel models have been proposed to describe non-WSS characteristics. However, these models cannot reflect the statistical variations in the size of the WSS interval of the real-time varying channels. Therefore, they are not suitable for serving V2V communication system design. In recent years, researchers have proposed different criteria to measure the size of the WSS interval in time-varying channels, as summarized below.

- Correlation Matrix Distance (CMD) criterion [8–10]: this criterion is used to measure whether the spatial structure of the channel undergoes significant changes and is applicable to multiple-input and multiple-output (MIMO) systems. References [11–14] apply CMD to analyze WSS interval characteristics in V2V MIMO channels.

- Spectral Divergence (SD) criterion [15]: this criterion is used to measure the absolute difference between spectral densities. The WSS interval characteristics in time-varying channels can be analyzed by comparing the differences in channel spectral densities at different time instants [5]. Reference [16] applies the SD criterion to study the WSS interval size in V2V time-varying channels, while reference

[17] analyzes the computational complexity of the SD criterion based on V2V channel measurements.

- Shadow Fading Correlation criterion [18]: this criterion is widely used to analyze the variations in statistical channel characteristics in traditional environments but has not been applied to V2V scenarios thus far. Shadow fading represents large-scale fluctuations of the received signal relative to the local mean. The spatial autocorrelation function of shadow fading describes the rate of change in local shadowing characteristics with the movement of mobile stations, and the corresponding shadow fading decorrelation distance is a measure of the rate of channel state change, which can be regarded as an estimate of the equivalent WSS interval size.

- Other criteria: Reference [19] proposes a local WSS interval definition method based on the Power Delay Profile (PDP) correlation, but it is not suitable for MIMO channels. Reference [20] improves the CMD criterion using a modified root-mean-square error criterion. Reference [21] estimates the interval of the WSS interval by comparing the variations in the wave number spectrum at different positions. Reference [22] measures the WSS interval by comparing the power spectral density of the delay at different time instants.

From the above references, it can be seen that there are limitations in the current research on scientific metrics for time-varying WSS intervals. In particular, there needs to be more research on the differences in estimation results between different criteria. Further research is required to scientifically and reasonably define the WSS interval.

In addition, the research on wireless propagation channels in NLOS scenarios in V2V communications has received increasing attention in recent years, and it can be classified into two main modeling methods: deterministic and statistical methods. Deterministic methods use detailed environmental information combined with Maxwell's equations (and their approximations) and specific boundary conditions of the environment to achieve deterministic predictions of wireless signal propagation. Deterministic modeling methods have been applied in V2V scenarios but mainly focused on LOS propagation scenarios [23–25]. Due to the complexity of NLOS scenarios (such as the specificity of large vehicle structures and randomness of buildings and trees around street crossings), traditional deterministic modeling techniques struggle to accurately reflect the real V2V channel and often have significant prediction errors [26]. Statistical methods, on the other hand, are based on actual measurements to model the statistical characteristics of wireless channel parameters and are widely used in V2V wireless channel research [27–31]. However, the measurement and modeling work for NLOS scenarios are relatively scarce. Existing works in this area can be summarized as follows.

4.1.1 Research on Large Vehicle Obstructing Scenarios in V2V Networks

Reference [32] studied the signal loss variation caused by truck blocking in the 11 GHz frequency band. Reference [33] investigated the propagation loss variation when multiple small vehicles block the communication link in the 60 GHz frequency band. However, the test frequency bands in these studies are different from the 5.9 GHz frequency band used in V2V networks, so they cannot be directly applied to the design of V2V communication systems. References [34,35] conducted channel measurements for V2V networks in NLOS scenarios in urban and highway scenarios. However, the NLOS propagation in these studies is mostly caused by small vehicles, and a certain proportion of LOS propagation is mixed in the measurements. As a result, the obtained results do not fully reflect the degradation of V2V link quality caused by large vehicle blocking. References [36,37] conducted measurements of the obstructing effect of truck bodies in static scenarios. However, these studies only performed field strength measurements at a few isolated locations, and the test results can only be used for qualitative analysis of the blocking effect of large vehicle bodies, but cannot establish a propagation model.

4.1.2 Research on Street Crossing Scenarios in V2V Networks

Reference [38] conducted V2V channel measurements in urban street crossing scenarios and found that buildings at street crossings have the greatest impact on wireless signal propagation loss. Reference [39] used channel data generated by a 3D ray tracing simulator to model the channels in urban street crossing scenarios. Reference [40] analyzed the V2V directional channel in urban street crossing scenarios based on measurements and pointed out that single bounce paths dominate the energy in this scenario. Reference [41] studied the maximum communication distance of V2V networks at urban street crossings. Reference [42] established path loss and fading models in urban street crossing scenarios based on measurements. However, the above studies mostly focused on urban street crossing scenarios and overlooked another important type of NLOS scenario, which is a suburban street crossing scenario. This limitation hinders the scientific design of V2V wireless communication systems in suburban scenarios.

The limitations of the aforementioned studies severely restrict the development of V2V channel simulation techniques [24,36] and hinder the further advancement of ITS.

Moreover, not only does the complex NLOS environment greatly degrade the quality of V2V wireless communication, but the time-varying and non-stationary characteristics of the V2V channel also significantly impact the performance of V2V wireless communication systems. In typical V2V propagation environments, the high-speed movement of vehicles and the random dynamic changes of scattering objects in the environment result in severe time-varying and non-stationary characteristics of the V2V wireless channel. Establishing a propagation model that incorporates the time-varying characteristics of the channel becomes crucial for V2V wireless channel measurement and modeling.

Due to the constantly changing distribution of multipath signals in a time-varying channel, it is challenging to statistically describe various parameters of the channel impulse response model. Traditional methods that focus on wireless channel parameters, such as statistical modeling of path loss, delay spread, and angle spread, only reflect the time-varying characteristics of the channel based on its external properties, which cannot accurately, realistically, and effectively simulate the variations in multipath structures in dynamic environments, nor can they thoroughly explain the multipath distribution and time-varying characteristics of the time-varying channel. Therefore, for modeling time-varying and non-stationary channels, traditional parameter-based modeling methods should not be used. Instead, more emphasis should be placed on tracking and describing the distribution and evolution characteristics of multipath components in the environment, such as the dynamic changes in the number of multipath, the multipath components' birth-death process, and the dynamic changes of multipath components within their own birth-death cycles. Accurate estimation and tracking of multipath signals in the V2V environment are key to describing the time-varying characteristics of multipath signals and forming the basis for the dynamic simulation of non-stationary V2V channels. Research on the time-varying characteristics of wireless channels in V2V scenarios contributes to the establishment of more realistic and accurate time-varying and non-stationary propagation models, serving the design of V2V wireless communication systems.

A significant amount of research has been conducted on V2V wireless channels based on measurements [4,27–29,43,44]. However, most of these studies have not specifically investigated the dynamic changes of multipath signals in V2V environments and instead assume that the parameters of the channel impulse response model remain constant over time. As a result, the findings of these studies cannot accurately reflect the true characteristics of time-varying V2V channels. References [11,13,16,17] have studied the definition of the WSS interval for time-varying V2V channels based on measurements. References [30,45] assume that the amplitude and delay of multipath signals in V2V channels do not change during their birth-death process and neglect the propagation characteristics of multipath signals in the angular domain. Currently, there is still a lack of in-depth analysis of the complete dynamic characteristics of multipath signals in V2V, such as the dynamic changes in the number of multipath, the birth-death process of multipath, and the dynamic changes in the amplitude, delay, and angle of multipath components. Further research is needed in this area.

Currently, there are relatively few research achievements in modeling time-varying channels (i.e., channel model parameters varying with time), and most of them focus on non-V2V scenarios. For example, reference [46] establishes an indoor directional channel simulator based on a two-state Markov model, which can be used to simulate the birth-death process of scattered signals. Reference [47] proposes a dynamic channel model for the 5.2 GHz indoor scenario based on a four-state Markov model. References [48,49] propose dynamic channel models for indoor scenarios based on ray-tracing simulators and use Poisson distribution to simulate the changes in the number of multipath in the environment. Reference [50] presents a dynamic time-varying model for the 1.8 GHz indoor channel. Reference [51] simulates the dynamic changes of multipath signals based on measurements and establishes a geometric-based statistical model. Reference [52] establishes a time-varying channel model for positioning based on geometric stochastic modeling theory and 5.2 GHz test data. The above research overview demonstrates that dynamic channel modeling has been widely applied to indoor propagation scenarios. Compared to V2V channels, indoor channels have fewer dynamic scattering objects, making it relatively easier to extract and track multipath signals. Although the measurement results from the above literature cannot be directly applied to V2V wireless channels, they provide valuable insights for modeling the dynamic and time-varying characteristics of V2V channels.

4.2 HIGHWAY TRAFFIC SCENARIO CHANNEL MEASUREMENT

To study the V2V wireless channel, field measurements are necessary as the data foundation for research. This section provides a detailed introduction to channel measurement in the V2V environment, including the measuring system, measuring scenarios, and preprocessing of measured data.

4.2.1 V2V Channel Measurement System

The selection of V2V testing systems is based on the requirements of practical modeling, and two sets of testing systems are mainly used. (1) Single Antenna Static Testing System: this system is based on a software-defined radio module and is used for static channel measurements in V2V scenarios where there are large vehicle obstructions. Due to considerations of highway traffic safety, it is difficult to conduct field measurements in such scenarios. It system allows for static measurements in these scenarios. (2) Multi-Antenna Dynamic Testing Platform: this platform is used for channel measurements in various V2V environments, including crossroad scenarios, traditional suburban, urban, and underground (tunnel-like) parking scenarios. It is equipped with multiple antennas and is capable of capturing dynamic channel variations in real time. The following sections will introduce these two testing platforms in detail.

The single antenna static testing system is based on the Wireless Open Access Research Platform (WARP) from Rice University in the United States [53]. The WARP wireless development board is used as the transmitter, and the measurement bandwidth is set to 15 MHz. This measurement bandwidth is larger than 10 MHz specified in the IEEE 802.11p V2V network standard [54], which meets the analysis requirements of V2V networks. Moreover, it effectively reduces the capture time for a complete signal, improving testing efficiency. Within the 15 MHz bandwidth, 385 equidistant samples were obtained. The transmission signal utilizes the Zadoff-Chu sequence [55] to ensure that the signal has a relatively flat magnitude response across the entire frequency range while maintaining sufficient transmit power. The length of the transmission sequence is set to 2^{10}.

In the static measurements, two distributed and synchronized WARP wireless development boards are used as transmitters for ease of result comparison (the arrangement of the two reflectors can be referred to in the subsequent description of V2V channel measurement scenarios). They operate at the transmission frequencies of 5.785 and 5.805 GHz, which are the highest frequencies supported by the WARP development boards. The transmission power is set to 28 dBm. At the receiver side, a Tektronix RSA5106A real-time spectrum analyzer is employed to capture the two complex signals in real time. The baseband filtering technique is used to separate the two signals, enabling dual-link synchronous measurements. Both the transmitter and receiver are equipped with 25 cm long L-com HG2458MGRD-RSP omni-directional antennas, which are placed on the top of the vehicles. Table 4.1 summarizes the core parameters of the single antenna static measuring platform.

Figure 4.1 illustrates the various modules of the static measurement system. During the measuring process, the WARP at the transmitter is set to continuous transmission mode.

TABLE 4.1   V2V Measurement System Parameters

Parameters	Static Measurement Platform Parameters	Dynamic Measurement Platform Parameters
Center Frequency	5.805 GHz	5.3 GHz
Transmitting Power	28 dBm	36 dBm
Measurement Bandwidth	15 MHz	60 MHz
Sampling Interval	2.5 cm	15 ms

(a) WARP demo board

(b) Tektronix RSA5106A real-time spectrum analyzer

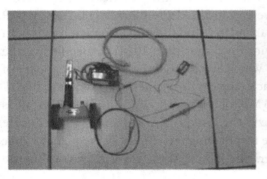

(c) Self made drive roller

(d) Calibration of measurements in an anechoic chamber

FIGURE 4.1   Static measurement system.

The spectrum analyzer at the receiver is connected to a homemade distance-triggered roller, with a distance-triggering accuracy of 0.1 cm. In the distance-triggered mode, the trigger interval is set to 0.5 times the wavelength, approximately 2.5 cm. The sampling interval of 0.5 times the wavelength ensures an adequate number of samples for analyzing large-scale channel characteristics and also guarantees sufficient accuracy for extracting small-scale parameters.

The multi-antenna dynamic measurement system utilizes the Aalto channel sounder [56] developed by Aalto University in Finland. This channel sounder is based on the principle of array switching, which enables the sequential capture of MIMO signals by rapidly switching between transmitter and receiver arrays. The Aalto channel sounder operates in the 5.3 GHz frequency band. Since wireless propagation characteristics do not exhibit significant changes within 10%–20% of the center carrier frequency [57], the analysis results at this frequency can be applied to 5.9 GHz V2V wireless communication systems. The channel sounder has a measuring bandwidth of 60 MHz and a sampling rate of 120 MHz, with a transmit power of 36 dBm. A pseudo-noise (PN) sequence with a length of 4.25 μs is used as the transmit signal in channel measurements. Capturing all MIMO antenna link signals takes approximately 1.632 ms. The channel sounder is set to trigger mode with a sampling interval of 15 ms. For a more detailed description of the Aalto channel sounder, please refer to the reference [56].

Figure 4.2 shows the antenna arrays of the dynamic MIMO measuring system. At the receiver side, a dual-polarized hemispherical antenna array is used. The hemispherical array consists of 15 dual-polarized (horizontal and vertical polarization) antenna elements, resulting in 30 signal reception links. The hemispherical antenna array can estimate the three-dimensional spatial angle characteristics of the wireless channel by observing the phase difference of the received signals in the horizontal and vertical planes. At the transmitter side, a uniform linear antenna array composed of four vertically polarized antennas is used. In channel measurements, both the transmitter and receiver antenna arrays are mounted on a wooden platform on the top of the vehicle.

(a) Receiver station hemispherical antenna array

(b) Transmitter uniform line antenna array

FIGURE 4.2 Dynamic MIMO measurement system antenna array.

4.2.2 V2V Channel Measurement Scenarios

This chapter describes V2V channel measurements conducted using both static and dynamic measurement systems. The former is employed to study the wireless channel propagation characteristics in V2V scenarios with large vehicle obstructions, typically found in urban street environments, while the latter is utilized to investigate the wireless channel characteristics in dynamic scenarios such as suburban areas, urban areas, underground (tunnel-like) parking, and street crossings within the V2V environment. The following section provides a detailed description of the V2V channel measurement scenarios.

The static urban street measurement was conducted on the campus of a university in the United States. To investigate the impact of large vehicle obstructions on V2V communication links and minimize the interference caused by dynamic changes in surrounding scatterers, a street block located far from the main roads within the university campus was selected for measuring. The measurements were conducted only on weekends when there was no passing traffic. In the static measurements, a standard model school bus and two small cars were used. Figure 4.3 illustrates the channel measurement setup in the static scenario. Transmitter 1 was placed on the roof of a small car, approximately 1.2 m away from the rear windshield. Transmitter 2 was placed at the front and rear edges of the school bus roof. Two different measure locations were selected for the static measurements, as shown in Figure 4.3b and c. Measurement Location 1 was an open area with no tall buildings around the school bus, making it difficult for multipath signals to reach the receiver through the scattering from surrounding objects. Measurement Location 2 was a more enclosed area where the school bus was surrounded by tall buildings, allowing multipath signals to reach the receiver not only through diffraction over the bus roof and sides but also through scattering from surrounding objects.

The static measurements were divided into LOS and NLOS scenarios. In the NLOS measurements, the school bus was positioned between the two small cars, as shown in Figure 4.3, in a stationary state. Transmitter 1 was placed at a predetermined position d_2, while the receiver moved at a constant speed along the d_1 trajectory. The measured "Car-to-Car" (CC) link under the obstruction of the large vehicle body was referred to as the NLOS-CC link. In the LOS measurements, both the CC link and the "Bus-to-Car" (BC) link were measured at the same street location, denoted as the LOS-CC and LOS-BC links, respectively. Transmitter 2 was placed at the front and rear edges of the school bus roof for the LOS-BC measurements, as shown in Figure 4.3b and c. It should be noted that at Location 1, the LOS-BC link was partially obstructed by the front edge of the school bus roof. The LOS measurements were conducted to compare with the NLOS measurements, allowing for a more accurate analysis of the impact of the large vehicle (i.e., school bus) obstruction on the wireless channel. Additionally, the LOS measurements helped discuss the placement of communication relay nodes on the bus roof and their effect on the performance of V2V NLOS wireless links. To ensure sufficient sampling accuracy, the driving speed was controlled at 1m/s to provide the trigger wheel with enough response time. In both Location 1 and Location 2 of the static measurements, different initial values were selected for transmitter 1's initial position (i.e., d_2). The detailed parameters and link names are summarized in Table 4.2. In the measurements, the range of d_1 was set to 300–400 m.

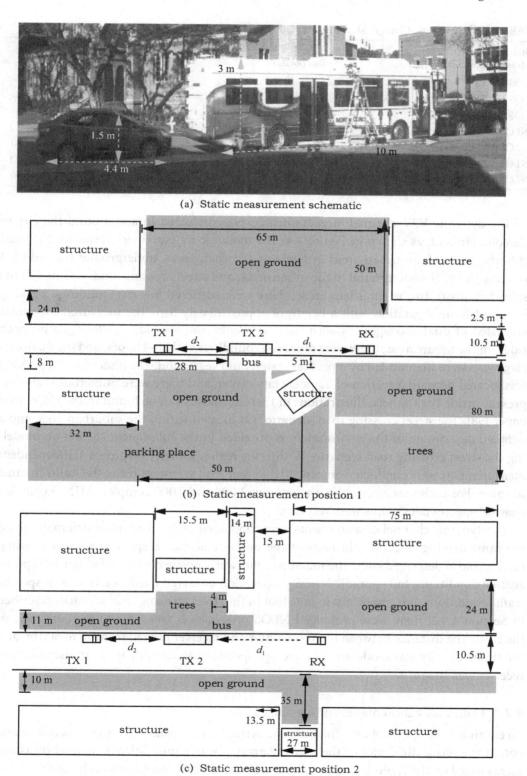

(a) Static measurement schematic

(b) Static measurement position 1

(c) Static measurement position 2

FIGURE 4.3 V2V Static channel measurement scenarios.

TABLE 4.2 V2V Static Channel Measurement Link Parameters

Static Channel Measurement Link	Bus Location	Transmitter	d_2/m
LOS-CC	—	—	—
LOS-BC-Location 1	Location 1	Transmitter2	—
LOS-BC-Location 2	Location 2	Transmitter 2	—
NLOS-CC-Location 1-1	Location 1	Transmitter 1	0.1
NLOS-CC-Location 1-2	Location 1	Transmitter 1	50
NLOS-CC-Location 2-1	Location 2	Transmitter 1	0.1
NLOS-CC-Location 2-2	Location 2	Transmitter 1	50

The dynamic V2V channel measurements were conducted in and around the city of Tapiola, Finland, as shown in Figure 4.4. The dynamic measurement scenarios included suburban areas (non-urban areas in Figure 4.4), urban areas, underground (tunnel-like) parking (located underground in the urban area), and street crossing roads (located in the suburban area). In the suburban areas, there were scattered low-rise buildings, parking lots, and some vegetation with a height of approximately 5 m. The measurement routes consisted of dual carriageways with wide sidewalks and sporadic traffic signs on both sides. In the urban areas, there were numerous buildings with 3–4 floors, and the dual-carriageways were surrounded by more traffic signs and vegetation. The underground parking was located beneath a section of Tapiola's city center, and there were numerous scatterers present inside the tunnels. Illustrations of these scenarios can be found in the referenced paper [58]. The street crossing road scenario was located within the suburban area, and a detailed description of the environment is provided in the subsequent section on modeling the street crossing road scenario. Within the region shown in Figure 4.4, independent measurements were conducted on over 30 different road segments using the Aalto channel sounder. For each measurement path, between 2,000 and 7,000 complete MIMO complex channel spectra data points were recorded.

The dynamic channel measurements were conducted using two cars, which maintained the same driving direction throughout the measurements, except in the street crossing road scenario described later. The measurements were mostly conducted under LOS propagation conditions, but sporadic NLOS propagation occurred due to variations in specific traffic conditions. It is important to note that in the street crossing road scenario described in Section 4.4.2, there were prolonged NLOS propagation areas due to the curvature of the road. The distance between the transmitter and receiver ranged from 10 to 500 m, and the traffic density was moderate. The average speed of the vehicles at the transmitter and receiver was 10 km/h.

4.2.3 Data Calibration and Preprocessing

To eliminate the effects of antenna gain and system components in channel modeling and obtain the phase difference of the antenna array (for dynamic MIMO channel measurements based on the Aalto University channel sounder), it is necessary to calibrate the measurement antennas in an anechoic chamber prior to the actual channel measurements, as

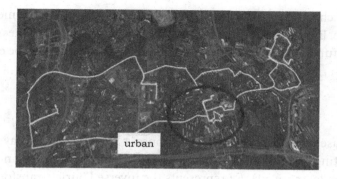

FIGURE 4.4   V2V Dynamic channel measurement environment.

shown in Figure 4.1d. The frequency-domain received signal in the actual channel measurement can be represented as

$$S_{Rx}(d,f) = S_{Tx}(f) H_{Tx}(f) H(d,f) H_{Rx}(f) \tag{4.1}$$

Where f is the frequency, d is the distance between the transmit and receive antennas (in the case of time-triggered mode in dynamic measurements, d can be replaced with the corresponding time t), $S_{Tx}(f)$ and $S_{Rx}(d,f)$ are the frequency domain responses of the transmit and receive signals, $H(d,f)$ is the frequency domain transfer function of the wireless channel, and $H_{Tx}(f)$ and $H_{Rx}(f)$ are the frequency domain transfer function of the system link (including antennas, system equipment links, cables, and other components). In antenna calibration, in order to eliminate the influence of $H_{Tx}(f)$ and $H_{Rx}(f)$ on channel modeling, the same system parameters (frequency, amplifier gain, and cable loss) as in the actual measurements are used for the calibration in the anechoic chamber. In the calibration measurement inside the anechoic chamber, the distance between the transmit and receive antennas is set to one meter, and the received frequency domain signal can be represented as

$$S_{ref}(f) = S_{Tx}(d=1,f) H_{Tx}(f) H_{ref}(d=1,f) H_{Rx}(f) \tag{4.2}$$

Dividing equation (4.1) by (4.2), we can obtain the frequency domain transfer function of the wireless channel in the actual measurement

$$H(d,f) = \frac{S_{Rx}(d,f)}{S_{ref}(d=1,f)} H_{ref}(d=1,f) \tag{4.3}$$

Where $H_{ref}(d=1,f)$ stands for the wireless channel frequency response in free space, with a propagation distance of 1 m. In the calibration of the dynamic MIMO measuring system, a similar antenna calibration is performed in an anechoic chamber to obtain the phase difference of different links in the MIMO antenna array for angle spectrum estimation in channel modeling. The calibration method and detailed description of the Aalto University channel sounder can be found in reference [56].

In subsequent chapters, channel modeling is primarily based on the measured channel impulse response. By performing an inverse Fourier transform on the measured frequency domain transfer function, the baseband channel impulse response can be obtained, as represented below

$$h(m,n,d,\tau) = F^{-1}\left[H(m,n,d,f)\right] \tag{4.4}$$

Where h is the baseband channel impulse response, τ represents the delay; m and n denote the m_{th} transmitting antenna and the m_{th} receiving antenna (for single-antenna systems in static measurements, $m = n = 1$), F^{-1} represents the inverse Fourier transform. In the aforementioned computation, a Hanning window is employed to filter the signal in order to mitigate the sidelobe effect [59]. The channel's transient PDP can be obtained from equation (4.4) and is defined as follows

$$P(m,n,d,\tau) = \left|h(m,n,d,\tau)\right|^2 \tag{4.5}$$

In the communication link corresponding to the m_{th} transmitting antenna and the m_{th} receiving antenna, the transient channel gain can be obtained using equation (4.6)

$$P_G(m,n,d) = \sum_{p=1}^{N_\tau} P(m,n,d,\tau_p) \tag{4.6}$$

Where N_τ is the number of sample points in each measured complete MIMO complex channel spectrum. It is important to note that in the time-triggered mode used by the dynamic measuring system, the distance d in equations (4.4)–(4.6) should be replaced with time t. The aforementioned transient channel impulse response, transient PDP, and transient channel gain will be utilized for the analysis and modeling of channel parameters in subsequent chapters.

4.3 DYNAMIC MODELING OF TIME-VARYING NON-STATIONARY CHANNELS

4.3.1 Methods for Measuring Non-Stationary Characteristics of Time-Varying Channels

Due to the dynamic variations of scatterers and the high-speed movement of transmitting and receiving nodes in V2V wireless channels, the wireless channel parameters exhibit time-varying and non-stationary characteristics. In order to quantitatively measure the changing features of wireless channel parameters in a statistically meaningful way, it is necessary to define a local window for analyzing V2V wireless channel parameters, known as the local WSS interval. Only the channel parameters extracted within the WSS interval have statistical significance and physical meaning. The scientific definition of the WSS interval is crucial for the analysis and modeling of wireless channels. The following section will provide a detailed explanation of the methodology for defining the WSS interval in V2V networks.

4.3.1.1 The Correlation Matrix Distance (CMD) Criterion

The CMD criterion, first proposed in reference [8], is a matrix-based decision method that effectively estimates whether significant changes have occurred in the spatial structure of the channel, such as multipath signal arrival and departure angles. The CMD criterion is widely used for estimating the size of the WSS interval in MIMO channel measurements due to its suitability for MIMO systems.

Considering a time-continuous MIMO measuring system with a signal sampling interval of t_{rep}, at any given time it_{rep}, the channel correlation matrix $R(it_{rep})$ can be defined at both the transmitter and receiver end to characterize the time-varying nature of the channel at different terminals. Alternatively, the complete MIMO channel matrix can be used to represent the global time-varying characteristics of the channel. The estimation methods for the aforementioned three definitions of correlation matrices are as follows [9]

$$R_{TX}(it_{rep}) = \frac{1}{W} \sum_{k=1}^{i+W-1} H(kt_{rep})^{T} H(kt_{rep})^{*}$$

$$R_{RX}(it_{rep}) = \frac{1}{W} \sum_{k=1}^{i+W-1} H(kt_{rep}) H(kt_{rep})^{H} \qquad (4.7)$$

$$R_{Full}(it_{rep}) = \frac{1}{W} \sum_{k=1}^{i+W-1} \text{vec}\left[H(kt_{rep}) \right] \cdot \text{vec}\left[H(kt_{rep}) \right]^{H}$$

Where $(\cdot)^{T}$ denotes matrix transpose, $(\cdot)^{*}$ represents matrix conjugate, $(\cdot)^{H}$ is the Hermitian transpose of the matrix, and the symbol vec[·] refers to stacking the matrix in column-wise order to form a column vector. In equation (4.7), W is an equivalent time-averaging window corresponding to 40 times the wavelength, used to filter out the impact of small-scale fading on the estimation results. $H(kt_{rep})$ is the complex channel matrix of the (equivalent) narrowband MIMO system at time kt_{rep} and is defined as follows

$$H(kt_{rep}) = \sum_{p=1}^{N_{\tau}} h(kt_{rep}, \tau_p) \qquad (4.8)$$

Where $h(kt_{rep}, \tau_p)$ is an $N_{TX} \times N_{RX}$ dimensional complex matrix composed of the channel impulse responses $h(kt_{rep}, \tau_p)$ [57]. N_{τ} is the number of time-domain test samples.

At any two time instances it_{rep} and jt_{rep}, the CMD definition for the channel correlation matrices $R(it_{rep})$ and $R(jt_{rep})$ is defined as follows

$$d_{corr}(i,j) = 1 - \frac{\text{tr}\left\{ R(it_{rep}) \cdot R(jt_{rep}) \right\}}{R(it_{rep})_{f} \cdot R(jt_{rep})_{f}} \qquad (4.9)$$

Where tr{·} denotes the trace of a matrix and $\|\cdot\|_{f}$ represents the Frobenius norm of a matrix. When two matrices are uncorrelated, the CMD equals to 1, and when two matrices are linearly correlated, the CMD equals to 0. Figure 4.5 shows the estimated two-dimensional time-varying CMD based on the complete MIMO channel matrix in a V2V suburban environment. Two distinct WSS regions (areas with smaller CMD values) are evident in Figure 4.5: 10–25 seconds and 70–95 seconds.

In the CMD criterion, a threshold value c_{th} can be selected to measure the magnitude of differences in the spatial structure characteristics of the channel at different times and locations. In this book, the WSS time window T is defined as the maximum time interval in which the CMD is smaller than c_{th}. It is important to note that, due to the time-varying nature of multipath channels, the WSS time window changes as the transmitter and receiver move. Therefore, the WSS time window T is a time-varying parameter and can be represented as

$$T(i) = \left(i'_{max} - i'_{min}\right) \cdot t_{rep} \tag{4.10}$$

where the minimum and maximum bounds of the time-varying WSS interval at time it_{rep} can be represented as [58]

$$\begin{cases} i'_{min} = \underset{0 \leq j \leq i-1}{\mathrm{argmax}} \, d_{corr}\left(i, j\right) \geq c_{th} \\ i'_{max} = \underset{i+1 \leq j \leq N_t - W}{\mathrm{argmin}} \, d_{corr}\left(i, j\right) \geq c_{th} \end{cases} \tag{4.11}$$

Where $i = 0, 1, \cdots N_t - W$, N_t is the total number of measured broadband signals.

The performance of the CMD criterion is greatly influenced by the threshold value c_{th}. To avoid the negative impact of subjective judgment differences on the selection of the threshold value under different test conditions, it is necessary to vertically compare the

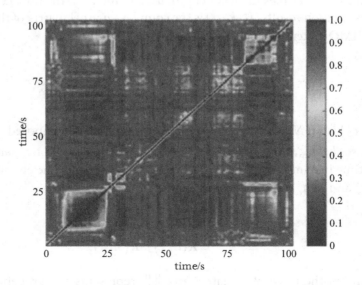

FIGURE 4.5 Example of CMD estimation in V2V suburban environment.

estimation performance corresponding to different threshold values and use a unified threshold value in the analysis. A larger threshold value results in a larger estimated WSS interval. Practical tests have shown [12] that there is a critical threshold value c'_{th}, where for smaller threshold values $c_{th} < c'_{th}$, the CMD estimation results do not change significantly, while for larger threshold values $c_{th} > c'_{th}$, the CMD estimation results increase significantly.

Therefore, this book adopts a similar principle to reference [12] to find an appropriate CMD threshold value for V2V scenarios by comparing the estimation performance with different threshold values and searching for the critical threshold value c'_{th}. Figure 4.6 shows the variation characteristics of the estimated time-varying WSS time window based on CMD in a V2V suburban environment for $c_{th} = 0.1$, $c_{th} = 0.2$, and $c_{th} = 0.3$. The estimated WSS time windows in the 10–25 seconds and 70–95 seconds regions are larger and correspond to the black regions in Figure 4.5. Additionally, from Figure 4.6, it can be observed that the estimated WSS time windows are very close to each other for threshold values of 0.1 and 0.2, while for a threshold value of 0.3, the estimated WSS time window is significantly larger, particularly in the 10–25 seconds region. Therefore, a threshold value of 0.2 can be chosen as the CMD estimation threshold for this test result. Similar examinations were conducted on the remaining test data, and it was found that the threshold value of 0.2 can effectively apply to the other measurement results as well. Therefore, this book adopts it for the CMD criterion in estimating the WSS interval. It is worth noting that reference [58] also recommends using 0.2 as the threshold value for CMD estimation in V2V environments.

4.3.1.2 The Spectral Divergence (SD) Criterion

The SD criterion was initially proposed in reference [15] to measure the difference between two non-negative, non-normalized power spectral densities. A smaller SD value indicates a stronger correlation and less difference between the two power spectral densities. This

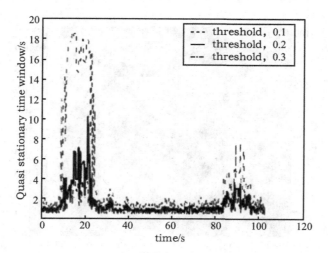

FIGURE 4.6 Example of estimating the optimal window size for CMD based on different threshold values in V2V suburban environment.

book combines SD with the PDP to describe the non-stationary characteristics of V2V time-varying channels. The SD between the instantaneous PDPs at any two-time instances it_{rep} and jt_{rep} is defined as follows

$$\gamma(i,j) = \ln\left(\frac{1}{N_\tau^2} \cdot \sum_{p=1}^{N_\tau} \frac{\tilde{P}(it_{rep},\tau_p)}{\tilde{P}(jt_{rep},\tau_p)} \cdot \sum_{p=1}^{N_\tau} \frac{\tilde{P}(jt_{rep},\tau_p)}{\tilde{P}(it_{rep},\tau_p)} \right) \tag{4.12}$$

Where $\tilde{P}$ represents the averaged PDP at the transmitter and receiver antenna arrays, and it can be expressed as

$$\tilde{P}(it_{rep},\tau_p) = \frac{1}{N_{TX}N_{RX}} \sum_{n=1}^{N_{TX}} \sum_{m=1}^{N_{RX}} \left| h(n,m,it_{rep},\tau_p) \right|^2 \tag{4.13}$$

Since the SD criterion is an unbounded estimate, it is possible for the estimated SD value to be arbitrarily large in certain cases. To facilitate analysis, this book normalizes the SD values estimated from any set of test data, such that the maximum SD value is equal to 1. Figure 4.7 illustrates the normalized two-dimensional time-varying SD estimated in a V2V suburban environment. Similar to Figure 4.5, there are two WSS regions within the intervals of 10–25 seconds and 70–95 seconds in Figure 4.7.

Similar to the CMD criterion, the SD criterion can also be used to measure the differences in channel PDP characteristics at different times and locations by selecting a threshold value c_{th}. By employing the same method as in equations (4.10) and (4.11), the WSS time window T can be defined as the maximum time interval where the SD is less than c_{th}. Comparing Figures 4.5 and 4.7, it can be observed that the estimated values using the SD criterion are much smaller than those using the CMD criterion. This indicates that the

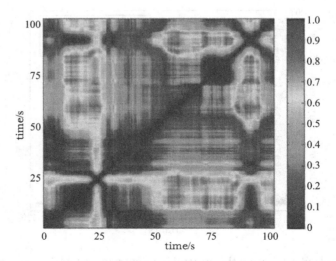

FIGURE 4.7   Example of normalized SD estimation in V2V suburban environment.

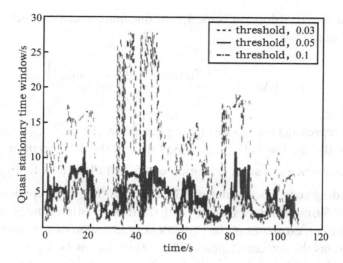

FIGURE 4.8 Example of estimating the optimal window size for normalized SD based on different threshold values in V2V suburban environment.

reasonable threshold value for the SD criterion will also be smaller than that for the CMD criterion. This book explores the non-stationary decision threshold for the SD criterion using the same approach as the CMD criterion. Figure 4.8 shows the variation characteristics of the estimated time-varying WSS window using the SD criterion in a V2V suburban scenario for c_{th} values of 0.03, 0.05, and 0.1. From Figure 4.8, it can be seen that when the threshold values are 0.03 and 0.05, the estimated WSS time windows are very close to each other, and both indicate significant WSS regions within the intervals of 10–25 seconds and 70–95 seconds (consistent with the estimation results of the CMD criterion). However, when the threshold value is set to 0.1, a larger WSS region appears within the interval of 30–70 seconds, which differs from the estimation results of the other two threshold values. This comparison demonstrates that the 0.1 threshold value introduces significant estimation errors. Therefore, this book selects 0.05 as the threshold value for SD estimation in this set of tests. By applying the same method to the remaining test data, it was found that the threshold value of 0.05 can effectively be applied to the other test results. Hence, this book adopts 0.05 as the threshold value for estimating the WSS window using the SD criterion.

4.3.1.3 The Shadow Fading Correlation Criterion

Since the variation of wireless channel statistics affects the autocorrelation characteristics of shadow fading, the autocorrelation coefficient of shadow fading can also serve as a measure of the WSS characteristics of the wireless channel [60]. Shadow fading is often quantified in terms of distance, and by dividing it by vehicle speed, the shadow fading samples can be obtained in the equivalent time domain. It should be noted that in the process of obtaining shadow fading samples in the time domain, a least squares (LS) regression fitting must be performed beforehand to remove path loss in the distance domain. As the V2V channel measurement system in this book adopts a time-triggered mode, the definition of shadow fading in the time domain is used, and the autocorrelation coefficient of shadow fading is defined as follows

$$\hat{\rho}_{\text{auto}}\left(\Delta it_{\text{rep}}\right) = \frac{E\left\{X_{\text{LS}}\left(it_{\text{rep}}\right)X_{\text{LS}}\left(it_{\text{rep}} + \Delta it_{\text{rep}}\right)\right\}}{\sigma_{\text{LS}}} \tag{4.14}$$

Where $X_{\text{LS}}\left(it_{\text{rep}}\right)$ represents the shadow fading component at time it_{rep}, and σ_{LS} is the standard deviation of the shadow fading. This book defines the decorrelation time as the relative time difference when the autocorrelation coefficient of shadow fading decreases to $\frac{1}{e}$. In the shadow fading correlation criterion, the decorrelation time represents the length of the equivalent WSS time window. By analyzing the decorrelation time of shadow fading in the V2V environment, estimates of the equivalent WSS region can be obtained.

Figure 4.9 presents the autocorrelation coefficients of shadow fading extracted from actual measurements in the V2V suburban environment. In the estimation of the path loss model, the relative distance between the transmitter and receiver is obtained by multiplying the propagation delay of the LOS signal in the PDP by the speed of light. Additionally, Figure 4.9 shows an exponential decay model for the autocorrelation coefficients of shadow fading [18]. It can be observed from Figure 4.9 that the exponential decay model aligns well with the test results in the V2V channel, with an equivalent decorrelation time of approximately 2.3 seconds.

4.3.1.4 Comparison
The three WSS time window decision methods summarized above have been widely used in the existing literature. However, there is currently a lack of cross-comparison among the results obtained from these three criteria, and the most suitable criterion for V2V

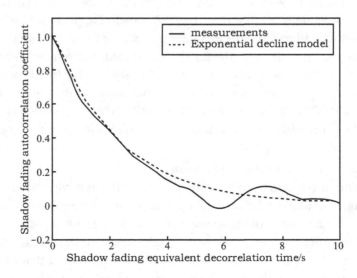

FIGURE 4.9   Example of shadow fading autocorrelation coefficient measurement in V2V suburban environment.

MIMO channel measurements remains to be studied. Therefore, in this section, dynamic measurement data is utilized to extract and compare the WSS time windows in the V2V environment using the three aforementioned criteria. Figures 4.10–4.12 present the cumulative distribution function (CDF) statistics of the estimated WSS time windows in three different scenarios: suburban, urban, and underground (tunnel-like) parking. Based on these results, a comparison and discussion of the following factors are conducted using the estimated WSS time window results.

4.3.1.4.1 Differences in WSS Time Estimation Using CMD Decision Criteria for Transmitter, Receiver, and Full MIMO Channel Matrix   From Figures 4.10–4.12, it can be observed that there are differences in the estimated WSS time using the CMD criterion for the transmitter, receiver, and full MIMO channel matrix (i.e., based on the calculation criterion of the three correlation matrices in equation 4.7). Specifically, the WSS time estimated using the transmitter matrix is larger by 5–20 seconds compared to the other two cases. This phenomenon differs from the conclusion drawn in Reference [13]. In Reference [13], the WSS time estimated using the CMD criterion for the transmitter and receiver was very close because the transmitter and receiver in Reference [13] employed the same antenna array. In the measurements conducted in this book, the size of the transmitter's four-element uniform linear antenna array is much smaller than the receiver's 30-element hemispherical antenna array. As a result, the spatial information (i.e., angular domain information) estimation accuracy at the receiver side is lower, leading to a larger estimated WSS time window. Furthermore, the estimated WSS time using the full MIMO channel matrix with the CMD criterion is slightly smaller than the results obtained using the receiver-side channel matrix with the CMD criterion, which indicates that using the former achieves the optimal performance under the CMD decision criterion.

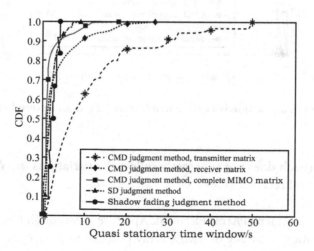

FIGURE 4.10   CDF of quasi-stationarity time window in V2V suburban environment.

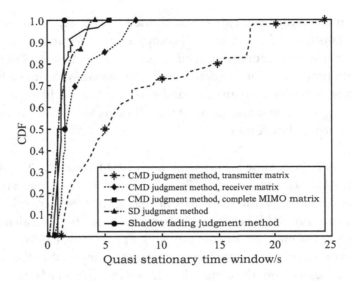

FIGURE 4.11   CDF of quasi-stationarity time window in V2V urban environment.

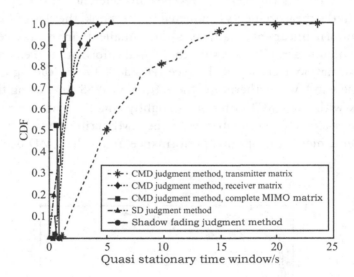

FIGURE 4.12   CDF of quasi-stationarity time window in V2V underground (tunnel-like) parking environment.

However, this approach does not separately reflect the variations in WSS characteristics at the transmitter and receiver sides.

4.3.1.4.2 Differences in Antenna Array Electronic Aperture   Although larger-scale arrays can improve the estimation accuracy of CMD, array processing theory states that the estimation accuracy in the angular domain depends on the electronic aperture of the antenna array and is independent of the number of array elements [61]. Therefore, in this section, based on the CMD decision criterion, we further examine the influence of differences in the electronic aperture of the antenna array on the accuracy of WSS window estimation. We consider all the receiver antennas and divide the transmitter antenna array into three groups to reconstruct three types of MIMO matrices:

- Type 1: using the first and second antennas of the transmitter.
- Type 2: using the first and fourth antennas of the transmitter.
- Type 3: using all four antennas of the transmitter.

Then, using the reconstructed MIMO matrices of the three types, we re-estimate the WSS time window under the CMD decision criterion. Figure 4.13 presents the results of the WSS time estimation based on the three types of MIMO matrices in the V2V suburban scenario using the CMD decision criterion. From Figure 4.13, it can be observed that the estimation result using the Type 1 MIMO matrix with a smaller electronic aperture is larger than the estimation result using the Type 2 MIMO matrix. This phenomenon aligns with the theoretical beamforming resolution: For a uniform linear array with M_{An} array elements and inter-element spacing Δd_{An}, the resolution of the signal's direction of arrival (DOA) $\Delta\varphi$ in the angular spectrum can be approximated as [61]

$$\Delta\varphi = \frac{\lambda}{(M_{An}-1)\cdot \Delta d_{An}} \cdot \frac{1}{|\cos(\varphi)|} \qquad (4.15)$$

Among them, φ represents an arbitrary angle value. Equation (4.15) indicates that the Type 2 MIMO matrix has better angular resolution compared to Type 1, resulting in a smaller estimated WSS time window. It should be noted that Equation (4.15) shows that the Type 2

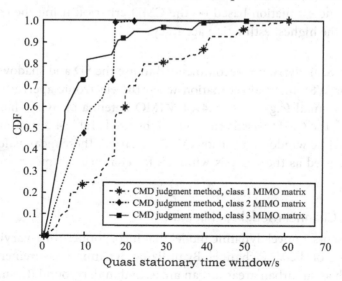

FIGURE 4.13 The influence of antenna array electronic aperture on quasi-stationarity time estimation.

and Type 3 MIMO matrices have the same angular resolution. However, the inter-element spacing Δd_{An} for the Type 2 MIMO matrix is larger than $\frac{\lambda}{2}$, which leads to potential ambiguity in the estimation [62]. Therefore, the Type 3 MIMO matrix achieves the highest estimation accuracy, which is consistent with the results shown in Figure 4.13. Based on the above comparison, it can be concluded that a larger electronic aperture of the antenna array improves the estimation accuracy of the WSS window.

4.3.1.4.3 Environmental Differences   From the comparison in Figures 4.10–4.12, it can be observed that the underground parking scenario has the smallest WSS time window. This is because in such tunnel-like structures, there are numerous scattering objects (e.g., walls, ceilings, other parked vehicles, metallic ventilation facilities, etc.). Similarly, in the urban area scenario where there are more scattering objects compared to the suburban area scenario, the estimated WSS time window is also smaller.

4.3.1.4.4 Differences in the Three Criterion   From the comparison in Figures 4.10–4.12, it can be seen that the WSS time estimation results based on the SD criterion and the shadow fading correlation criterion are very close to the estimation results based on the CMD criterion using the complete MIMO matrix. Since the SD and shadow fading correlation criterion does not rely on the MIMO antenna system, the following conclusions can be drawn:

- The SD and shadow fading correlation criterion have good WSS time estimation accuracy and are not constrained by the electronic aperture of the antenna system.

- The CMD criterion can reflect the changes in WSS characteristics of the transmitter and receiver channels separately, but it requires a larger electronic aperture of the antenna array to ensure sufficient estimation accuracy.

- The WSS time estimation based on the CMD criterion using the complete MIMO matrix has the highest estimation accuracy.

Based on the above analysis, it is recommended to use the SD and shadow fading correlation criterion for WSS interval estimation when the electronic aperture of the antenna array is relatively small (e.g., 2×2 or 4×4 MIMO antenna array). When the electronic aperture is larger, the CMD-based criterion can be used. Table 4.3 summarizes the mean values $\bar{T}$ of WSS time windows in various V2V scenarios. These mean values of WSS time windows will be used as the analysis windows for channel parameter analysis in subsequent chapters.

4.3.2 Dynamic Channel Model Architecture

This section presents a novel dynamic model that incorporates time-varying non-stationary characteristics of the V2V channel. By utilizing dynamic measurement data collected in scenarios such as suburban areas, urban areas, and underground (tunnel-like) parking, the dynamic variations of multipath signals are modeled.

TABLE 4.3   Mean of Generalized Quasi-Stationarity Time Window in V2V Dynamic Measurements

Scenario	CMD Transmitting Metrics (seconds)	CMD Receiving Metrics (seconds)	CMD MIMO Metrics (seconds)	SD (seconds)	Shadowing Metrics (seconds)
Suburban	11.21	3.51	1.52	2.11	2.46
Urban	7.41	2.39	1.15	1.19	1.12
Underground (tunnel-like) Parking	6.23	1.63	0.95	1.60	1.37

The channel impulse response of the V2V dynamic wideband directional channel model can be represented as follows

$$h\left(t_i, \tau_l(t_i), \varphi_l(t_i)\right) = \sum_{l=1}^{N(t_i)} a_l(t_i) e^{j\varphi_l} \cdot \delta\left(\tau - \tau_l(t_i)\right) \cdot \delta\left(\varphi - \varphi_l(t_i)\right) \tag{4.16}$$

Where $\delta(\cdot)$ is the unit impulse function, $N(t_i)$ represents the total number of multipath signals at time t_i, and $a_l(t_i)$, $\tau_l(t_i)$, and $\varphi_l(t_i)$ denote the amplitude, delay, and angle of arrival of the l_{th} multipath signal at time t_i, respectively. In equation (4.16), $a_l(t_i)$, $\tau_l(t_i)$, and $\varphi_l(t_i)$ are all time-varying functions. φ_l represents the phase of the l_{th} multipath signal, which follows a uniform distribution within the range of $[0°, 360°]$.

In a time-varying V2V channel, the amplitude, delay, and angle of arrival of each multipath signal vary with time. Due to the random motion of mobile stations and the presence of numerous dynamic scatterers in the environment, each multipath component may randomly appear at one moment and randomly disappear at another moment. Within the birth and death process of a multipath signal, its amplitude, delay, and angle can undergo changes. To simulate a time-varying V2V channel, this book introduces the concept of multipath signal birth and death times to characterize the dynamic variations of the aforementioned parameters. For the sake of convenience in subsequent discussions, this book defines the following three sets of multipath signals.

- $L_{i \to i}$ represents the set of multipath signals that first appear at time t_i. Their index numbers can be represented as $l_{i \to i} = 1, 2, \cdots, N(t_{i \to i})$.

- $L_{j \to i}$ represents the set of multipath signals that first appear at time t_j and still exist at time t_i, where $0 < j < i$. Their index numbers can be represented as $l_{j \to i} = 1, 2, \cdots, N(t_{j \to i})$.

- L_i represents the set of all multipath signals that still exist at time t_i, which means $L_i = L_{1 \to i} \bigcup L_{2 \to i} \bigcup \cdots L_{j \to i} \bigcup \cdots L_{i-1 \to i} \bigcup L_{i \to i}$

By identifying and tracking all the sets $L_{i \to i}$ and $L_{j \to i}$ within the multipath signals, it is possible to reconstruct the directional channel impulse response at any given time. This allows for the incorporation of the time-varying characteristics of the channel. Therefore, the model in equation (4.16) can be further represented as follows

$$h\big(t_i,\tau(t_i),\varphi(t_i)\big)=\sum_{j=1}^{i-1}\sum_{l_{j\to i}=1}^{N(t_{j\to i})}\Big[a_{l_{j\to i}}(t_i)e^{j\varphi l_{j\to i}}\cdot\delta\big(\tau-\tau_{l_{j\to i}}(t_i)\big)\cdot\delta\big(\varphi-\varphi_{l_{j\to i}}(t_i)\big)\Big]+$$

$$\sum_{l_{j\to i}=1}^{N(t_{j\to i})}\Big[a_{l_{i\to i}}(t_i)e^{j\varphi l_{i\to i}}\cdot\delta\big(\tau-\tau_{l_{j\to i}}(t_i)\big)\cdot\delta\big(\varphi-\varphi_{l_{j\to i}}(t_i)\big)\Big]$$

(4.17)

Where

$$N(t_i)=N(t_{i\to i})+\sum_{j=1}^{i-1}N\big(t_{j\to i}\big)$$

(4.18)

The first term on the right-hand side of equation (4.17) represents the multipath signals that first appeared before time t_i, i.e., the old multipath signals. It is important to note that old multipath signals can fall into two categories: one is the multipath signals that continue to exist at subsequent times, and the other is the multipath signals whose life-cycle ends at time t_i. By assigning an initial birth and death time to each new multipath signal, the dynamic model can automatically capture the dynamic characteristics of both types of signals. The second term represents the multipath signals that first appear at time t_i, i.e., the new multipath signals. The key to dynamic channel modeling lies in integrating the evolution of multipath signals' birth and death at any given time. By separately modeling the dynamic evolution process of new and old multipath signals, it is possible to achieve the dynamic simulation of the channel impulse response at any given time.

For the time-varying V2V channel, the parameters in equation (4.16) vary with time. To scientifically simulate the variation process of these parameters from a statistical perspective, it is necessary to define a WSS region, also known as a WSS time window, and perform statistical modeling of the channel parameters within each WSS region. It is important to note that equation (4.17) shows that the lifecycle of a multipath signal is usually longer than the size of a single WSS region, meaning that a multipath signal can exist in different consecutive WSS regions. Therefore, the birth and death cycle of a multipath signal can be measured using WSS regions. The dynamic characteristics within the birth and death cycle of a multipath signal can be described from two aspects:

- Large-scale variations: the power, delay, and angle of a multipath signal exhibit changes across different WSS regions. This means that the aforementioned parameters remain constant within a specific WSS region but vary as the WSS region changes.

- Small-scale variations: the instantaneous power of a multipath signal follows a specific distribution within each WSS region, while the angle and delay remain constant.

In summary, in the proposed dynamic wideband directional channel model, the following parameters are modeled based on measurements:

- Multipath signal birth and death time.
- The number of newly appeared multipath signals in each WSS region.
- The positions (delay and angle) of newly appeared multipath signals in each WSS region.
- The initial power of each multipath signal when it first appears.
- The dynamic variations of power/delay/angle within the birth and death cycle of each multipath signal.
- Small-scale fading characteristics of multipath signals within each WSS region.

By dynamically and statistically modeling the above parameters, a time-varying dynamic channel model can be established to achieve dynamic simulation of the V2V channel.

4.3.3 Multipath Extraction and Dynamic Tracking

This section will provide a detailed explanation of the extraction and tracking of multipath signals in V2V channels, including the amplitude α, delay τ, and azimuth angle information of the signals φ. The extraction and tracking of the time-varying characteristics of multipath signals will facilitate the subsequent statistical modeling of V2V time-varying channels.

4.3.3.1 Multipath Signal Spatial-Temporal Spectrum Analysis and Extraction

In this section, the Bartlett beamforming method is used to analyze the time-varying angular spectrum $\overline{P_{\text{ang}}}$, where $\overline{P_{\text{ang}}}$ represents the spectrum function averaged over each WSS time window. Previous studies [62–65] have shown that the Bartlett angular spectrum exhibits high spatial analysis accuracy and low computational complexity when the dimension of the antenna array is large. In this chapter, we only consider the azimuth of arrival (AOA) φ dimension for the following reasons

- The number of antennas at the transmitting end of the dynamic test system is small, limiting the resolution of estimating the departure angles of multipath signals.
- In the measurement scenario, the distribution of scattering objects remains relatively constant in the elevation angle dimension. Therefore, only AOA information is considered in dynamic modeling.

By performing peak detection on the averaged Bartlett angular spectrum within each WSS region at time t and delay τ, the time, delay, and AOA information of the multipath signals can be obtained. In the angular spectrum, a signal is recorded as a useful multipath component only when the angular difference between multipath signals exceeds 30°, as the AOA resolution of the test system is 26° [56]. To better extract multipath signals, only co-polarization situations are considered, and a data processing threshold 20 dB lower than the strongest signal is set (as the effective testing range of the dynamic test system is 20 dB) to ensure that the extracted signals are valid multipath components.

Figure 4.14a shows the estimated logarithmic AOA-delay spectrum $\overline{P_{\text{ang}}}$ at a certain moment in the V2V suburban scenario. The spectrum function is normalized so that the maximum value is 0 dB. An AOA of 0° indicates that the multipath signals are incident from directly in front of the receiving antenna. Figure 4.14b shows the multipath signals detected from Figure 4.14a using the aforementioned method. A total of eight multipath signals are detected, with the LOS signal located at 240 ns and an AOA φ of 0°. From the comparison between Figure 4.14a and b, it can be observed that the stronger multipath signals are successfully detected.

4.3.3.2 Tracking the Dynamic Multipath Signal Birth-Death Process

To capture the dynamic variations of multipath signals in a time-varying channel and simulate the evolution of multipath signals' birth and death processes from a statistical perspective, the Multipath Component Distance (MCD) [66] is used to track the changes in multipath signals over consecutive WSS time intervals. MCD is a measure of the difference between multipath signals. A smaller MCD indicates a higher similarity between two multipath signals [67]. This characteristic of MCD can be used for multipath signal clustering [68] and tracking, as well as estimating their birth and death time (similar to tracking the centroid of multipath signal clusters using MCD as described in reference [69]).

For the convenience of expression, in the following text, i and $i+1$ represent the indices of two consecutive WSS time windows. For any two multipath signals l_1 and l_2 with consecutive WSS time window indices, their multipath information can be represented as follows

$$l_1 \in L_i : \left[\overline{a_1(t_l)}, \overline{\tau_1(t_l)}, \overline{\varphi_1(t_l)}\right],$$

$$l_1 \in L_{i+1} : \left[\overline{a_2(t_{l+1})}, \overline{\tau_2(t_{l+1})}, \overline{\varphi_2(t_{l+1})}\right]$$

(4.19)

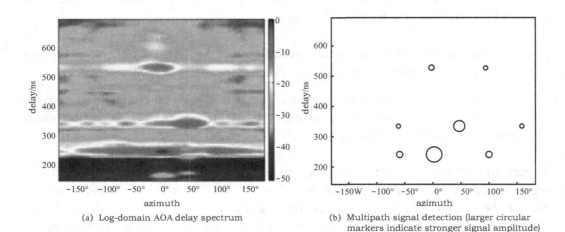

(a) Log-domain AOA delay spectrum

(b) Multipath signal detection (larger circular markers indicate stronger signal amplitude)

FIGURE 4.14 V2V Suburban scenario measurement example.

Based on the suggestion from reference [68], this book introduces a heuristic amplitude correction factor in the estimation of MCD. The modified expression for MCD estimation is as follows

$$\text{MCD} = \left(\frac{1}{\Delta a_{\max}} \left| \frac{\overline{a_1(t_l)}}{\overline{a_2(t_{l+1})}} \right| \right) \cdot \sqrt{\left\| \text{MCD}_{\text{AOA}} \right\|^2 + \text{MCD}_\tau^2} \qquad (4.20)$$

Where the estimated value of MCD in the AOA domain (as a vector) can be represented as follows

$$\text{MCD}_{\text{AOA}} = \frac{1}{2} \left\| \begin{pmatrix} \sin\left(\overline{\theta_1(t_l)}\right)\cos\left(\overline{\varphi_1(t_l)}\right) \\ \sin\left(\overline{\theta_1(t_l)}\right)\sin\left(\overline{\varphi_1(t_l)}\right) \\ \cos\left(\overline{\theta_1(t_l)}\right) \end{pmatrix} - \begin{pmatrix} \sin\left(\overline{\theta_2(t_{l+1})}\right)\cos\left(\overline{\varphi_2(t_{l+1})}\right) \\ \sin\left(\overline{\theta_2(t_{l+1})}\right)\sin\left(\overline{\varphi_2(t_{l+1})}\right) \\ \cos\left(\overline{\theta_2(t_{l+1})}\right) \end{pmatrix} \right\| \qquad (4.21)$$

Where $\left\| \text{MCD}_{\text{AOA}} \right\|$ represents the length of the vector mentioned above, $\overline{\theta_1(t_l)} = \overline{\theta_2(t_{l+1})} = 90°$. The estimated value of MCD in the delay domain can be represented as follows

$$\text{MCD}_\tau = \frac{\tau_{\text{std}}}{\left(\Delta\tau_{\max}\right)^2} \cdot \left| \overline{\tau_1(t_l)} - \overline{\tau_1(t_{l+1})} \right| \qquad (4.22)$$

Where

$$\Delta\tau_{\max} = \max\left(\overline{\tau_l}\right) - \min\left(\overline{\tau_l}\right)$$

$$\Delta a_{\max} = \frac{\max\left(\overline{a_l}\right)}{\min\left(\overline{a_l}\right)}, l \in \left(L_i \bigcup L_{i+1}\right) \qquad (4.23)$$

τ_{std} represents the standard deviation of the time delays within the set $l \in \left(L_i \bigcup L_{i+1}\right)$.

To assess the similarity of multipath signals in different WSS intervals using MCD and track the multipath signals, this book sets a reference threshold $\in$. When the MCD between two multipath signals in consecutive WSS intervals is below the threshold value $\in$, these two multipath signals are considered the same signal. This indicates that the multipath signal has persisted from the previous WSS interval to the next one, with possible changes in its amplitude, delay, and angle during this process. The multipath tracking algorithm in this book is based on the principle of bidirectional matching and is described as follows.

1. Calculating the MCD between any two multipath signals in L_i and L_{i+1} and storing the results in an $N(t_i) \times N(t_{i+1})$ dimensional MCD matrix $\boldsymbol{D}$.

2. If the following conditions are satisfied

$$D_{u,v} \leq \in$$

$$u = \overset{\text{argmin}}{u} \left(D_{u \in N(t_i), v} \right) \tag{4.24}$$

$$u = \overset{\text{argmin}}{v} \left(D_{u, v \in N(t_{i+1})} \right)$$

The u_{th} multipath at time t_i and the v_{th} multipath at time t_{i+1} are considered the same multipath signal. Consequently, each of these two multipath can be assigned a unique identification code to facilitate their pairing.

3. Examine all multipath signals at time t_i and t_{i+1}, and use equation (4.24) to match the multipath signals that meet the conditions.

4. Calculate the MCD between any two multipath signals in L_{i+1} and L_{i+2}, and repeat steps ① and ②. If the w_{th} multipath signal at time t_{i+2} matches the v_{th} multipath signal at time t_{i+1}, then the w_{th} multipath signal at time t_{i+2} inherits the identity code of the v_{th} multipath signal at time t_{i+1}. This process continues iteratively.

5. Repeat the above steps for signals at time t_{i+2} and onwards. Perform bidirectional matching between every two consecutive WSS intervals and assign identity codes to all multipath signals based on specific criterion and matching results.

The multipath tracking algorithm starts from $i = 1$. By grouping and classifying multipath signals with the same identity code, information about their birth, death, and evolution can be obtained. It is important to note that in the tracking and modeling process, multipath signals that disappear and reappear are ignored. When a multipath signal disappears within a WSS interval, its lifecycle is considered to end. This assumption simplifies channel modeling without significantly degrading the model's accuracy. The MCD threshold is set to a specific value $\in = 0.5$, which has been shown in practical measurements to effectively track multipath signals and has been adopted in reference [69]. Prior to tracking multipath signals based on measurement results, the amplitudes, delays, and azimuth angles of all multipath signals are normalized with respect to the LOS signal. This normalization ensures that the LOS signal has a power of 0 dB, a delay of 0 ns, and an azimuth angle of 0°. Normalization mitigates the impact of changes in the relative position of the mobile station in dynamic measurements on the model parameters.

Figure 4.15 shows an example of multipath signal tracking in a V2V suburban scenario. The results of multipath tracking, indicated by LS linear regression curves, are shown in

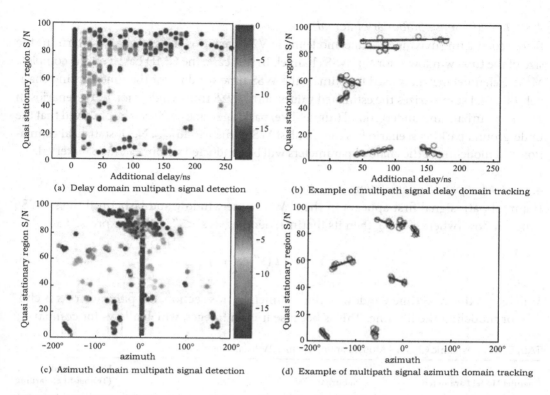

(a) Delay domain multipath signal detection

(b) Example of multipath signal delay domain tracking

(c) Azimuth domain multipath signal detection

(d) Example of multipath signal azimuth domain tracking

FIGURE 4.15 Simplified representation of multipath signals in V2V suburban scenario measurements.

Figure 4.15b and d (the LOS path is not included for clarity). Multipath signals with a lifespan of fewer than six times the WSS time window (as defined in the subsequent text) are not depicted in Figure 4.15b and d for simplicity. However, these short-lived multipath signals are still considered in the modeling process. From Figure 4.15, it can be observed that the proposed multipath tracking algorithm takes into account the variations in both the delay domain and the angle domain, resulting in high accuracy. Only when multipath signals are similar in both delay and angle domains, the tracking algorithm groups them together as the same multipath signal. Furthermore, the variations in the delay domain and the angle domain of multipath signals are independent of each other, and the evolutionary trajectory of multipath signals during their lifespan (i.e., the slope of the linear regression curve) is unrelated to the initial delay and angle values. This phenomenon aligns with the physical characteristics of dynamic V2V channels, where the distribution of scatterers in the environment exhibits dynamic randomness.

4.3.4 Time-Varying Channel Modeling

The goal of this section is to utilize the extracted and tracked results of the multipath signals from the previous section to select appropriate statistical distribution functions for modeling the dynamic variations of multipath signal parameters. This will enable us to statistically model the dynamic changes of multipath signals.

4.3.4.1 Wide-Sense Stationary Interval

Before conducting dynamic channel modeling in V2V scenarios, it is essential to determine the size of the time window $\bar{T}$ for the WSS channel. In this case, the CMD based on the complete MIMO channel matrix is used to estimate the WSS time window for the time-varying channel. Table 4.4 summarizes the estimated values of the WSS time window for V2V scenarios in suburban, urban, and underground (tunnel-like) parking scenarios. It can be observed that the underground parking scenario has the smallest WSS time window $\bar{T}$. Next, statistical estimation and modeling of the channel parameters will be conducted within each WSS interval.

4.3.4.2 Multipath Signal Lifetime

If a multipath signal first appears in the s' WSS time window and lasts until the s'' WSS time window (where $s' < s''$), then its lifetime, denoted as $s' < s''$, can be expressed as:

$$\tilde{T} = \bar{T} \cdot \Delta s = \bar{T}(s'' - s' + 1) \qquad (4.25)$$

Where $\bar{T}$ is the WSS time window of the channel. In this section, the parameter Δs is chosen for modeling the lifetime. This is because it is an integer, which allows for comparing

TABLE 4.4 Dynamic Channel Model Parameters in V2V Scenarios

Channel Model Parameters	Suburban	Urban	Underground (Tunnel-like) Parking
$\bar{T}$	1.52 seconds	1.15 seconds	0.95 seconds
s_{max}	49	48	29
$\mu_{\Delta s}$	1	1	1
$\sigma_{\Delta s}$	10.5	13.0	6.8
p_{11}	1.30	1.81	2.39
p_{12}	7.70	3.05	3.10
P_{21}	1.22	1.48	1.61
P_{22}	9.76	9.12	2.82
σ_j	46.9°	76.2°	104.1°
$\tau_{a,bp}$	91.67 ns	91.67 ns	66.67 ns
q_1	−0.19	−0.19	−0.30
q_2	−0.008	−0.009	−0.005
q_3	−12.80	−12.93	−13.54
$k_\tau, (k_\tau > 0)$	9.05	8.09	2.89
$k_\tau, (k_\tau < 0)$	−5.15	−6.84	−2.53
$k_\varphi, (k_\varphi > 0)$	5.19	6.00	17.58
$k_\varphi, (k_\varphi < 0)$	−3.46	−7.54	−9.77
$k_a, (k_a > 0, k_\tau > 0)$	0.060	0.060	0.018
$k_a, (k_a < 0, k_\tau < 0)$	−0.045	−0.063	−0.051
$k_a, (k_a > 0, k_\tau < 0)$	0.027	−0.042	−0.068
$k_a, (k_a < 0, k_\tau < 0)$	−0.035	−0.041	−0.031
$k_a, (k_a > 0, k_\tau = 0)$	0.009	0.049	0.016
$k_a, (k_a < 0, k_\tau = 0)$	−0.006	−0.004	−0.014
$\tau_{m,bp}$	16 ns	8 ns	8 ns
m_1	3.31	1.20	1.06
m_2	0.53	0.30	0.35

its distribution characteristics with the Poisson distribution. In the modeling process of Δs, the impact of the LOS path is not considered since the LOS path exists throughout the entire test route.

Figure 4.16a presents the Probability Density Function (PDF) fitting results based on the measurement value of Δs. The performance of Poisson, Gamma, Exponential, and truncated Gaussian distributions fitted using the LS method is compared in the figure. It can be observed that the truncated Gaussian distribution exhibits the best-fitting performance. Additionally, Figure 4.16a shows that the Exponential distribution is not suitable for describing the characteristics of Δs. Therefore, in the dynamic channel modeling presented in this book, a modeling approach based on Markov random processes is not employed. This is because the prerequisite for modeling based on Markov random processes is that the lifetimes of multipath signals follow an Exponential distribution [46–48]. Finally, a truncated Gaussian distribution with thresholds within the range $[1, s_{max}]1$ is used as the model for Δs. The mean of the truncated Gaussian distribution is $\mu_{\Delta s}$, the standard deviation is $\sigma_{\Delta s}$, and s_{max} represents the maximum value obtained from measured Δs.

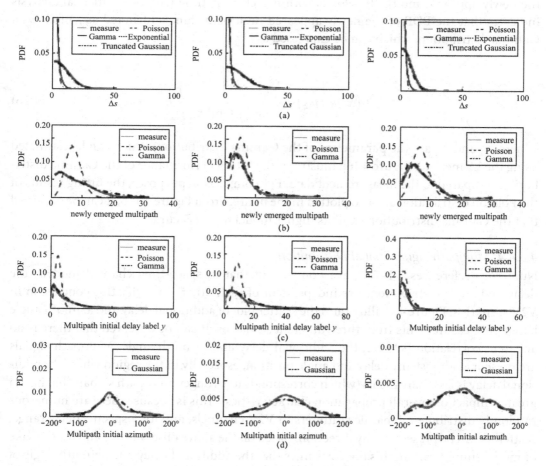

FIGURE 4.16   PDF fitting results of measurement data (left to right: suburban, urban and underground (tunnel-like) parking).

Table 4.4 summarizes the model parameters of the truncated Gaussian distribution for different scenarios. It can be observed that the $s_{\max}$ and $\sigma_{\Delta s}$ values are the smallest for the underground parking scenario, indicating that the lifetime of multipath signals in this environment is relatively short due to the presence of numerous scatterers in the underground parking environment.

4.3.4.3 Numbers of Multipath Signals

From equation (4.18), the total number of multipath signals $N(t_i)$ at time t_i is determined by the combined contribution of newly appeared multipaths, denoted as $N(t_{i \to i})$, and the existing multipaths, denoted as $N(t_{j \to i})$, where $0 < j < i$. Since $N(t_{j \to i})$ depends on the $N(t_j)$ multipaths present at time t_j and their lifetimes, in this case, it is only necessary to model the number of newly appeared multipaths within each WSS interval (where the conclusion from Section 4.3.3 is used to assign an initial lifetime to each newly generated multipath signal). This allows for the dynamic simulation of time-varying channels.

By tracking the multipath signals in the measurement results, it is possible to identify the newly appeared multipath signals within each WSS time window. Statistical analysis indicates that the PDF of the newly appeared multipath signals (denoted as f_1) follows a Gamma distribution, which can be expressed as

$$f_1\left(x; p_{11}, p_{12}\right) = \frac{x^{p_{11}-1} \cdot \exp\left(-\dfrac{x}{p_{12}}\right)}{\left(p_{12}^{\,p_{11}}\right) \cdot \mathrm{Gm}\left(p_{11}\right)} \tag{4.26}$$

Where p_{11} and p_{12} are the parameters of the Gamma distribution, which can be estimated using the LS method. Figure 4.16b illustrates the fitting performance of the Gamma distribution compared to the measurement data. For comparison purposes, the fitting results of the Poisson distribution are also plotted in the figure. From Figure 4.16b, it can be observed that the Gamma distribution exhibits higher model fitting accuracy.

4.3.4.4 Multipath Signal Spatial Distribution

Numerous references [71–73] have pointed out that in space-time channel models, the delay and angle distributions are independent of each other. To verify this conclusion in V2V scenarios, Figure 4.17 illustrates the scatter plot of additional delay and azimuth angle based on measurements from three different scenarios. It can be observed that there is no apparent correlation between the additional delay and azimuth angle. Multipath signals are present at almost any delay and angle position, especially in the region where the additional delay is less than 400 ns (which corresponds to stronger multipath signals that have a greater impact on overall propagation characteristics). This is because there are numerous randomly distributed dynamic scatterers in V2V channels, and for a specific delay range, scatterers can be present in any direction. Based on the above observations and for the sake of model simplification, this book assumes that the additional delay and azimuth angle of multipath signals are independent of each other. It is important to note that this phenomenon differs from the measurement results of indoor channels mentioned in reference [74].

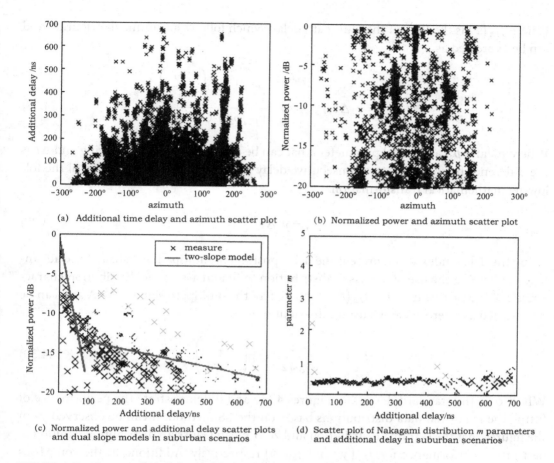

(a) Additional time delay and azimuth scatter plot

(b) Normalized power and azimuth scatter plot

(c) Normalized power and additional delay scatter plots and dual slope models in suburban scenarios

(d) Scatter plot of Nakagami distribution m parameters and additional delay in suburban scenarios

FIGURE 4.17　Scatter plot of channel parameters in V2V scenario.

Reference [74] states that in indoor LOS channels, multipath signals with larger additional delays have a smaller fluctuation range in their direction angles. This is because multipath signals with larger additional delays in indoor LOS scenarios mostly come from backscattering off the rear walls, and these backscattered waves have similar AOA to the LOS path. In contrast, in indoor NLOS scenarios described in reference [74], where scatterers can be present in any direction, the correlation between delay and angle is reduced, which is similar to the measurement results.

In this section, modeling is conducted only for the positions (i.e., delay and AOA) of newly generated multipath signals. This is because the positions of existing multipath signals can be estimated using the initial multipath position information and an evolving model based on the position information within the multipath signal's lifetime (detailed information will be discussed later). The PDF of the position information for newly generated multipath signals within each WSS interval can be represented as

$$f_2(\tau,\varphi) = f_{2,\tau}(\tau) \cdot f_{2,\varphi}(\varphi) \tag{4.27}$$

Where $f_{2,\tau}(\tau)$ is the PDF of the multipath delay, which follows a Gamma distribution and can be expressed as

$$f_{2,\tau}(\tau = y \cdot \Delta \tau; p_{21}, p_{21}) = \frac{y^{p_{21}-1} \cdot \exp\left(-\dfrac{y}{p_{22}}\right)}{(p_{22} p_{21}) \cdot \mathrm{Gm}(p_{21})} \tag{4.28}$$

Where p_{21} and p_{22} are model parameters that can be obtained through LS fitting and $\Delta \tau$ is the difference between any two consecutive delay samples. Here, with the help of the following mathematical transformation

$$\tau = y \cdot \Delta \tau \tag{4.29}$$

Using the delay index y to represent the delay position of multipath signals has the advantage of allowing the use of a Poisson distribution to fit and assess the distribution characteristics of integer values of y. $f_{2,\phi}(\phi)$ is the PDF of the multipath signal's AOA and can be represented as a zero-mean Gaussian distribution:

$$f_{2,\phi}(\varphi) = \frac{1}{\sigma_\varphi \sqrt{2\pi}} \exp\left(-\frac{\varphi^2}{2\sigma_\varphi^2}\right) \tag{4.30}$$

Where σ is the standard deviation. Figures 4.16c and d demonstrate the performance of fitting the measured data distributions based on the LS method. It can be observed from the figures that the Gamma distribution and zero-mean Gaussian distribution exhibit the best fitting performance for $f_{2,\tau}(y)$ and $f_{2,\phi}(\phi)$ respectively. Additionally, the von Mises distribution [75,76] was also examined for $f_{2,\phi}(\phi)$, but due to its poor fitting performance, it is not shown in Figure 4.16d. Table 4.4 summarizes the estimation results of parameters p_{11}, p_{22} and σ_φ for the three V2V scenarios, where the underground parking scenario has the largest σ_φ.

4.3.4.5 Initial Power of Multipath Signals

For a multipath signal that first appears in the s'_{th} WSS time window and lasts until the s''_{th} WSS time window, its initial power can be represented as

$$l: \left[\overline{a_l(t_{s'})}, \overline{\tau_l(t_{s'})}, \overline{\varphi_l(t_{s'})}\right] \tag{4.31}$$

The variations in the amplitude, delay, and azimuth angle of the multipath signal within its own lifetime can be simulated as a function of its lifetime index $s(s' \le s \le s'')$ and its initial state information $\left[\overline{a_l(t_{s'})}, \overline{\tau_l(t_{s'})}, \overline{\varphi_l(t_{s'})}\right]$. In this section, we will first establish a model for $\overline{a_l(t_{s'})}$ using the test data.

Figure 4.17b displays the scatter plot of the normalized power of multipath signals versus the AOA in the three V2V scenarios. It can be observed from the figure that there are strong multipath signals at any AOA. This indicates that the amplitude of the multipath signal is independent of the AOA. Therefore, in this case, $\overline{a_l(t_{s'})}$ is modeled as a function of the additional delay. To achieve a higher fitting accuracy and a smaller model complexity, this book adopts a dual-slope model to model $\overline{a_l(t_{s'})}$ in the logarithmic domain, which can be represented as

$$20\log\left(\overline{a_l(t_{s'})}\right) = \begin{cases} q_1 \cdot \tau[\text{ns}], 0 \leq \tau < \tau_{a,bp} \\ q_2 \cdot \tau[\text{ns}] + q_3, \tau_{a,bp} \leq \tau \leq 700 \text{ ns} \end{cases} \tag{4.32}$$

Where $\tau_{a,bp}$ is the breakpoint of the dual-slope model, and q_1, q_2 and q_3 are model parameters that can be obtained through LS fitting. Figure 4.17c displays the fitting performance of the dual-slope model in the V2V suburban scenario, demonstrating that the dual-slope model can effectively capture the relationship between multipath normalized power and additional delay. The estimated results of parameters q_1, q_2 and q_3 and τ_{bp} for each scenario are provided in Table 4.4.

4.3.4.6 Evolution of Multipath Signals

Continuing the approach from Section 4.3.3, we will model the variations in the amplitude, delay, and AOA of multipath signals within their own lifetimes. During the modeling process, we differentiate between two types of multipath signals: long-lifetime signals and short-lifetime signals, denoted as

$$\Delta s = s'' - s' + 1 : \begin{cases} \geq 6, \text{long lifetime signals} \\ < 6, \text{short lifetime signals} \end{cases} \tag{4.33}$$

Where the threshold value of 6 is determined based on statistical fitting of the test data, as analysis of the test duration indicates that the fitting performance of the proposed linear regression model is poor when the multipath signal's Δs is less than 6.

To avoid poor fitting performance, for short-lifetime signals, their amplitude, delay, and AOA are assumed to be constant throughout their lifetime, represented as

$$\overline{a_l(t_s)} = \overline{a_l(t_{s'})}$$

$$\overline{\tau_l(t_s)} = \overline{\tau_l(t_{s'})} \tag{4.34}$$

$$\overline{\varphi_l(t_s)} = \overline{\varphi_l(t_{s'})}$$

where $s' \leq s \leq s''$. It is important to note that in this case, the initial information of the multipath signal is used as the evolution model for short-lifetime signals, rather than the statistical mean within their lifetime. This is because in dynamic channel simulations, it is necessary to assign initial amplitudes, delays, and AOAs to each multipath signal beforehand. According to the aforementioned assumption, the amplitude, delay, and AOA of short-lifetime signals remain constant throughout their lifetime.

For long-lifetime signals, a simple linear polynomial function [46,47] is used to model the variations of multipath delay and AOA during their own birth and death process. It can be represented as

$$\overline{\tau_l(t_s)} = \overline{\tau_l(t_{s'})} + k_\tau(s - s' + 1)$$

$$\overline{\varphi_l(t_s)} = \overline{\varphi_l(t_{s'})} + k_\varphi(s - s' + 1)$$

(4.35)

Where k_τ and k_φ are model parameters that can be estimated through LS fitting. Measurement results indicate that k_τ and k_φ are independent of each other and independent of the specific signal's delay and AOA, as shown in Figure 4.15. This implies that within a multipath signal's birth and death time, k_τ and k_φ can take positive, negative, or zero values, and these cases occur with equal probability. Therefore, this book adopts an equiprobable event simulation of the evolution process of multipath delay and AOA, that is

$$P_r(k_\tau > 0) = P_r(k_\tau < 0) = P_r(k_\tau = 0) = \frac{1}{3}$$

$$P_r(k_\varphi > 0) = P_r(k_\varphi < 0) = P_r(k_\varphi = 0) = \frac{1}{3}$$

(4.36)

Where $P_r(\cdot)$ represents probability. In the aforementioned equiprobable events, the mean values of k_τ and k_φ obtained based on measurement results are taken as the final model parameters, denoted as $\left(k_{\tau,(k_\tau > 0)}, k_{\tau,(k_\tau < 0)}, k_{\tau,(k_\tau = 0)}\right)$ and $\left(k_{\varphi,(k_\varphi > 0)}, k_{\varphi,(k_\varphi < 0)}, k_{\varphi,(k_\varphi = 0)}\right)$ respectively. Table 4.4 summarizes the measurement results of these parameters in different scenarios, where $k_{\tau,(k_\tau = 0)} = k_{\varphi,(k_\varphi = 0)} = 0$.

For modeling the variation of the power of the multipath signal $\overline{a_l(t_s)}$ within its lifetime, reference [72] suggests using a smooth monotonic function, while the variation of the signal power over its lifetime is assumed to follow a sinusoidal function. However, these methods are not consistent with the measurement results. To simplify the model, this book still uses a linear polynomial function to simulate the variation of $a_l(t_s)$, represented as

$$\overline{a_l(t_s)} = \overline{a_l(t_{s'})} + k_a(s - s' + 1)$$

(4.37)

k_a is a model parameter. Since the amplitude of the multipath signal is independent of the AOA, k_a and k_φ are also independent of each other. Furthermore, measurements have shown that for different values of k_τ, k_a can be either positive or negative, and they occur with equal probability. Therefore, the evolution process of the multipath signal power within its lifetime (k_τ) can also be modeled as an equiprobable random variable, given by:

$$P_r(k_a > 0, k_\tau > 0) = P_r(k_a < 0, k_\tau > 0) = P_r(k_a > 0, k_\tau < 0) = P_r(k_a < 0, k_\tau < 0)$$

$$= P_r(k_a > 0, k_\tau = 0) = P_r(k_a < 0, k_\tau = 0) = \frac{1}{6}$$

(4.38)

According to the measurement results, the mean value of k_a under the aforementioned equiprobable events can be obtained. It can be represented using the following symbols

$$
\begin{bmatrix}
P_r\left(k_a > 0, k_\tau > 0\right) \\
P_r\left(k_a \langle 0, k_\tau \rangle 0\right) \\
P_r\left(k_a > 0, k_\tau < 0\right) \\
P_r\left(k_a < 0, k_\tau < 0\right) \\
P_r\left(k_a > 0, k_\tau = 0\right) \\
P_r\left(k_a < 0, k_\tau = 0\right)
\end{bmatrix}
\tag{4.39}
$$

The aforementioned parameters are summarized in Table 4.4. Please note that in the measurement results of this book, the case where $k_a = 0$ did not occur.

In summary, this book employs equiprobable random variables to simulate the evolution process of long-lived multipath signal amplitudes, delays, and AOA within their lifetimes. For each class of events corresponding to the equiprobable random variables, a linear model based on the birth-death time index, denoted as s, is used to describe them.

4.3.4.7 Small-Scale Fading Characteristics of Multipath Signals

To describe the small-scale fading characteristics of a certain multipath signal $\left[a_l(t_s), \tau_l(t_s), \varphi_l(t_s)\right]$ in $\overline{P}_{\text{ang}}(t_s, \tau_s, \varphi_s)$ work, the following steps are taken. Firstly, within the s_{th} WSS region, the corresponding multipath signals are extracted based on the transient delay-angle spectrum using the parameter $\left[\tau_l(t_s), \varphi_l(t_s)\right]$. Then, the average amplitude of the extracted multipath signals is subtracted in each WSS time window. Finally, the small-scale distribution fitting and analysis of the amplitude $a_l(t_s)$ of the multipath signal are performed. The number of samples available for small-scale parameter estimation in each WSS time window is given by

$$
N_{ss} = \frac{\overline{T}}{t_{\text{rep}}}
\tag{4.40}
$$

Similar to the previous discussion, four common small-scale fading distributions, namely Ricean, Rayleigh, Nakagami and Weibull, are considered for V2V channels. A combined approach using the AIC and the KS at a 95% confidence level is employed to select the optimal small-scale distribution function. The test results indicate that the Nakagami distribution exhibits the best fitting characteristics in over 85% of the cases. To describe the small-scale fading characteristics more accurately, this book proposes a Nakagami distribution parameter m model for multipath signals based on measurements. As mentioned earlier, the amplitude of multipath signals, denoted as φ, is independent of each other, and therefore, the Nakagami parameter m is also independent of φ. Figure 4.17d illustrates a scatter plot of m against additional delay in a suburban scenario. It can be observed that within a small range of additional delay, the value of m is relatively large, which is consistent with the measurement results in reference [77]. After 16 ns, the value of m drops below 0.6. The measurement results in urban and underground parking scenarios exhibit similar trends to Figure 4.17d. Therefore, a piecewise constant function is employed here to describe the variation of m as follows

$$
m = \begin{cases} m_1, \tau < \tau_{m,bp} \\ m_2, \tau \geq \tau_{m,bp} \end{cases} \tag{4.41}
$$

Here, $\tau_{m,bp}$ is the breakpoint of the m model, and m_1 and m_2 are the estimated mean values of m in different regions. The results of the aforementioned parameters are summarized in Table 4.4. It can be observed from the table that multipath signals with larger delays exhibit more severe small-scale fading characteristics compared to Rayleigh fading. Additionally, in the underground parking lot/tunnel scenario, multipath signals with smaller delays exhibit small-scale fading characteristics similar to Rayleigh fading.

4.3.5 Dynamic Model Simulation and Validation

This section will introduce the usage of the aforementioned dynamic model and validate its performance in time-varying dynamic V2V channels based on measurement data.

4.3.5.1 Steps for Dynamic Model Simulation

Figure 4.18 provides a detailed description of the simulation process for the V2V dynamic model.

The specific steps are as follows:

Step 1 Start by setting the initial number of multipaths and the total number of WSS regions required. The birth-death time, initial position, and initial power of the multipaths are generated by the proposed model. After initialization, discard multipath signals with power lower than 20 dB below the strongest signal.

Step 2 In the second WSS region, first check the birth-death time of the existing multipath signals to determine whether to retain or discard them. For the multipath signals that should be retained in the second WSS region, they are divided into two groups: long-lifetime signals and short-lifetime signals. The short-lifetime signals inherit their power, delay, and AOA from the previous WSS region. For the long-lifetime signals, their power, delay, and AOA information is updated based on the proposed model. To determine k_φ, generate an equiprobable random variable with values 1, 2, or 3, where each value corresponds to one equiprobable event of k_φ in equation (4.36). Then update the AOA of the multipath using k_φ and equation (4.35). Next, generate an equiprobable random variable with values 1, 2, 3, 4, 5 or 6 to determine k_a and k_τ. Each value corresponds to one equiprobable event of k_a and k_τ in equation (4.38). Then update the amplitude and delay of the multipath using k_a, k_τ and equation (4.37).

Step 3 Generate N_2 new multipath signals in the second WSS region. If $N_2 > 0$, assign birth-death time, initial position, and initial power to these N_2 new multipath signals. Finally, discard multipath signals in the old set where the power is 20 dB below the strongest signal.

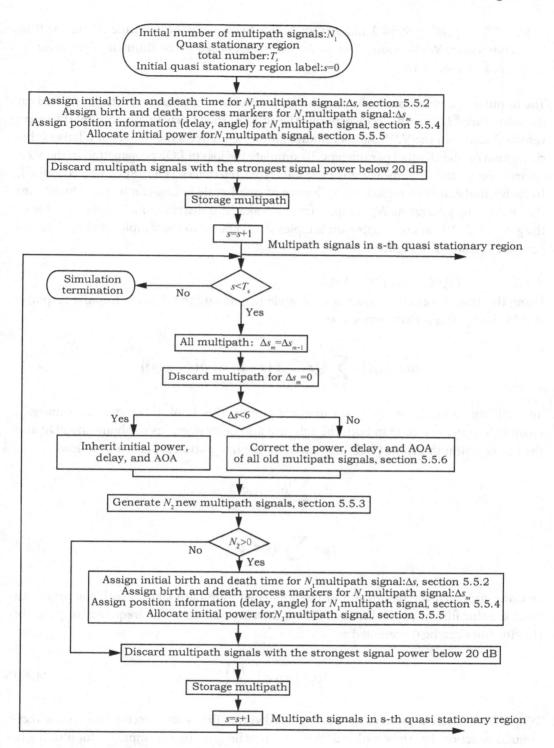

FIGURE 4.18   V2V Dynamic model simulation process.

Step 4 By repeating Steps 2 and 3, it is possible to simulate the dynamic channel within consecutive WSS regions. The model parameters used in the simulation are summarized in Table 4.4.

The impulse response obtained using the aforementioned simulation method is based on the normalized LOS path. To incorporate the effect of the LOS path, one can set the inter-vehicle distance in the V2V channel simulation, which reflects the impact of relative vehicle motion on the channel parameters. To simulate the loss of LOS propagation in the V2V environment, statistical measurement results from references [4,43,78] can be consulted. In each simulated WSS region, the influence of small-scale fading can be introduced into the channel by generating N_{ss} samples from a Nakagami distribution. The mean value of the generated Nakagami distribution samples should be set to the sample amplitude values generated in Step 2 of Figure 4.18.

4.3.5.2 Model Performance Validation

Using the aforementioned model, it is possible to generate the channel impulse response within the s_{th} WSS region, denoted as

$$\tilde{h}\left(s,\tilde{\tau}_l,\tilde{\phi}_l\right)=\sum_{l=1}^{N(s)}\tilde{a}_l(s)^{e^{j\tilde{\phi}_l}}\cdot\delta\left(\tau-\tilde{\tau}_l(s)\right)\cdot\delta\left(\phi-\tilde{\phi}_l(s)\right) \tag{4.42}$$

To facilitate comparison with measurement results, this book integrates the simulated channel impulse response in both the azimuth and delay domains to obtain the PDP and the Power Azimuth Profile (PAP), respectively. The integration process is as follows

$$P=\sum_{\varphi}\left|\tilde{h}\left(s,\tilde{\tau}_l,\tilde{\phi}_l\right)\right|^2$$

$$P_A=\sum_{\tau}\left|\tilde{h}\left(s,\tilde{\tau}_l,\tilde{\phi}_l\right)\right|^2 \tag{4.43}$$

In order to match the bandwidth of the simulated multipath signals with the measured results, a sinc filter is applied to the PDP for filtering purposes. The frequency response of the sinc filter can be represented as:

$$H(f)=\text{rect}\left(\frac{f}{2B}\right) \tag{4.44}$$

Where $B=60\,\text{MHz}$ represents the testing bandwidth. The symbol rect(·) denotes the rectangular function. For the simulated PAP, a Bartlett beamformer is employed for filtering in order to obtain the angular spectrum corresponding to the measured results.

Using the proposed model, this book simulated the normalized channel impulse response within 100 consecutive WSS regions in a suburban scenario. In the simulation, the initial number of multipaths was set to 7, which represents the statistical mean of newly generated multipath signals within each WSS region in the suburban environment. To reduce complexity, the simulation did not include small-scale fading characteristics, as they need to be filtered out in the model validation process of PDP and PAP. Figure 4.19 presents the comparison between the measured and simulated results of PDP and PAP in the V2V suburban scenario. It can be observed from the figure that the measured and simulated results are closely aligned. However, it is important to note that the model comparison in Figure 4.19 is based on randomly generated channels with statistical properties, which limits the analysis of the agreement between the measured and simulated results to a qualitative level [79].

To further validate the proposed dynamic model, the validation of second-order statistical properties is performed using the previously defined Root Mean Square (RMS) delay spread τ_{rms} and RMS azimuth spread φ_{rms}. Figure 4.20 illustrates the comparison of Cumulative Distribution Function (CDF) of the second-order statistical properties

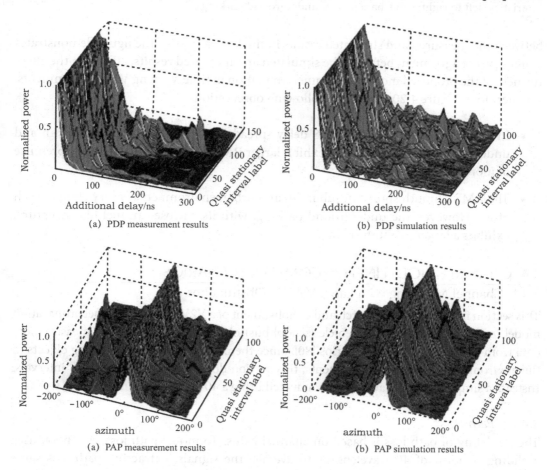

(a) PDP measurement results

(b) PDP simulation results

(a) PAP measurement results

(b) PAP simulation results

FIGURE 4.19 Comparison of V2V suburban environment measurement and simulation results.

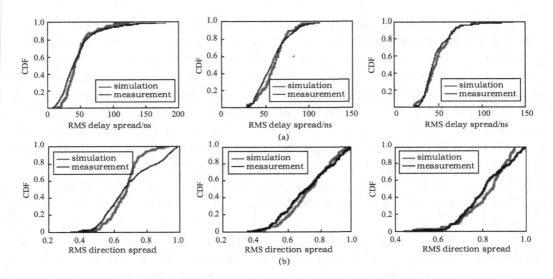

FIGURE 4.20 Comparison of channel measurements and simulation second-order statistical characteristics (left to right: suburban, urban, underground parking).

between the measured and simulated results in the V2V scenario. The figure demonstrates a high level of agreement between the simulated and measured results, affirming the effectiveness of the proposed model for channel simulation in time-varying V2V environments. Furthermore, Figure 4.20 reveals the following observations

- In suburban scenarios, the RMS delay spread is minimal. In contrast, urban and underground parking scenario exhibit larger RMS delay spreads due to the presence of more scattering objects.

- The RMS azimuth spread in suburban and urban environments is very close to each other. However, the underground parking, with its enclosed tunnel-like structure, exhibits a larger azimuth spread.

4.4 CHANNEL MODELING OF COMPLEX SCENARIOS

4.4.1 Channel Modeling with Large Vehicle Obstructions

This section introduces the research and establishment of a V2V wireless channel propagation model in the presence of large vehicle (school bus) obstructions, using static measurement data from urban street block scenarios [80]. Since there are no dynamic scattering objects in the environment of static measurements, this section does not estimate the WSS intervals. Instead, a simple statistical analysis is conducted using a 40-wavelength interval [81].

4.4.1.1 Path Loss

The modeling of path loss is based on channel gains. To model path loss, this book uses a sliding window of 40 wavelengths to average the signals, extracting path loss samples that contain shadow fading information. Since the static testing system adopts a distance-triggered mode, statistical modeling of path loss can be performed based on distance. In static measurements, the distance between the transmitting and receiving antennas can be represented as:

$$d = \begin{cases} d_1 + d_2 + L_{\text{bus}} + L_{\text{car}}, \text{NLOS} \\ d_1 + L_{\text{bus}} + 1.2, \text{LOS} - \text{BC} - \text{Location1} \\ d_1 + 3.2, \text{LOS} - \text{BC} - \text{Location2} \end{cases} \quad (4.45)$$

Where d_1 and d_2 are the distances between the bus and the cars on both sides; $L = 10.0\,\text{m}$ and $L_{\text{car}} = 4.4\,\text{m}$ represent the length of the bus and the cars, respectively. The distances of 1.2 and 3.2 m correspond to the distances between the rear and front of the test car and the receiving antenna, respectively.

Firstly, the observed path loss in V2V static LOS and NLOS scenarios is examined. Figures 4.21 and 4.22 depict the variations of measured path loss with respect to the distance between the transmitter and receiver antennas in V2V static LOS and NLOS scenarios, respectively. Several phenomena can be observed from these figures.

1. The measured results of path loss in LOS scenarios generally align with the dual-slope model [57]. This observation is consistent with the findings in reference [82]. For the link "LOS-BC-Position 1," when the receiver is close to the bus (where LOS is obstructed by the top of the bus), the path loss approximately follows a simple single-peaked diffraction model. Additionally, the path loss in LOS scenarios is lower by 15–20 dB compared to NLOS scenarios. This difference is higher by 6–10 dB compared to the measurement results in reference [34] and approximately 11 dB higher than the recommended values in reference [26]. It is important to note that, compared

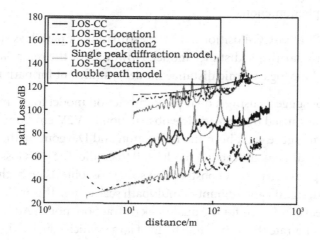

FIGURE 4.21 Relationship between path loss and distance between transmitter and receiver antennas in static V2V LOS scenario. (For clarity, the overall path loss for the link "LOS-BC-Location1" has been increased by 30 dB, and the overall path loss for the link "LOS-BC-Location2" has been decreased by 30 dB.)

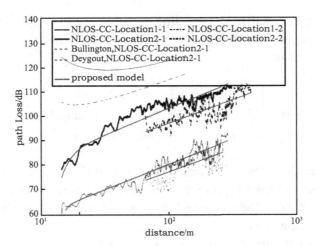

FIGURE 4.22 Relationship between path loss and distance between transmitter and receiver antennas in static V2V NLOS scenario. (For clarity, the overall path loss for links "NLOS-CC-Location1-1" and "NLOS-CC-Location1-2" has been reduced by 20 dB.)

to the path loss results in this book, references [26,34] mainly focus on NLOS scenarios caused by obstruction from surrounding small vehicles, and specifically consider situations where the communication terminal is far from the obstructing objects.

2. In NLOS scenarios, in the case of a fixed distance between the transmitting and receiving antennas, smaller values of d_2 correspond to larger path loss. This observation aligns with the diffraction mechanism of wave propagation in shadowed regions, where diffraction loss increases as d_2 decreases. However, traditional path loss models such as the single-slope logarithmic distance model [83] are unable to reflect this propagation phenomenon.

3. For the link "NLOS-CC-Position 2-1," the estimation of path loss should be based on a multi-slope logarithmic distance model. However, for the remaining NLOS scenarios, a single-slope logarithmic distance model is sufficient for path loss estimation.

4. Reference [26] suggests using a dual-peak diffraction model to estimate the additional diffraction loss caused by large vehicle obstruction in V2V environments. However, as observed from Figure 4.22, both the Bullington and Deygout methods [57], which are based on the dual-peak diffraction model, fail to predict the path loss accurately. This is because in typical urban street obstruction scenarios, objects on both sides of the street contribute to a significant amount of multipath scattering. This makes it challenging to satisfy the ideal conditions for the dual-peak diffraction model. Additionally, multipath signals may penetrate the front windshield of large vehicles such as buses with relatively small loss. However, for metal trucks and large trailers, this penetration loss tends to be more significant, resulting in more severe shadowing effects on V2V communication links. Therefore, a multi-peak diffraction model may be more applicable to wireless channels under the obstructions of trucks and large trailers.

The analysis above indicates that in NLOS scenarios, the presence of surrounding scattered objects in urban street environments has a significant impact on wireless propagation loss. Simplified geometric optical models are inadequate to accurately predict path loss in such scenarios. Additionally, in the presence of large vehicle obstructions, it is necessary to establish a statistical path loss model that reflects the influence of d_1 and d_2 on propagation loss, as shown in Figure 4.3. In this case, traditional distance-based power-law models are no longer applicable.

Next, let's focus on modeling path loss in LOS scenarios. For the links "LOS-CC" and "LOS-BC-Location 2," the classic dual-slope model can be used, and the path loss PL follows the same expression as equation (4.46).

$$\mathrm{PL_{Two-Ray}} = \left(\frac{\lambda}{4\pi}\right)^2 \left| \frac{e^{-jk_{d2}d_d}}{d_d} + \varepsilon \frac{e^{-jk_{d2}d_r}}{d_r} \right|^2 \tag{4.46}$$

In the dual-slope model, the wireless signal undergoes coherent and destructive interference between two multipath signals, resulting in oscillations in the path loss curve, as shown in Figure 4.21. From Figure 4.21, it can be observed that the path loss for the links "LOS-CC" and "LOS-BC-Location 2" approximately follows the path loss curve of the dual-slope model.

To model the path loss for the link "LOS-BC-Location 1," it is necessary to differentiate between the dual-path propagation region and the diffraction region. In this case, the breakpoints for these two propagation regions with respect to the distance from the bus can be expressed as follows

$$d_b = \left(\frac{h_{\mathrm{bus}} - h_{\mathrm{car}}}{h_{\mathrm{antenna}}}\right) \cdot L_{\mathrm{bus}} \tag{4.47}$$

Where $h_{\mathrm{bus}} = 3.0\,\mathrm{m}$, $h_{\mathrm{car}} = 1.5\,\mathrm{m}$, and $h_{\mathrm{antenna}} = 25\,\mathrm{m}$. From this, we can calculate $d_b = 50\,\mathrm{m}$. From Figure 4.21, it can be observed that the path loss for the link "LOS-BC-Location 1" follows dual-path propagation after the breakpoint distance d_b, and it follows a single-lobe diffraction model before the breakpoint distance d_b. In the single-lobe diffraction model, the obstacle can be approximated as an infinitely large electromagnetic absorber plane located vertically between the transmitting and receiving antennas. The additional loss caused by the obstacle, denoted as J_{dB}, can be expressed as [84]

$$J_{\mathrm{dB}}(v) = -20\lg\left(\frac{\sqrt{\left|1 - C(v) - S(v)\right|^2 + \left|C(v) - S(v)\right|^2}}{2}\right) \tag{4.48}$$

Where $C(v)$ and $S(v)$ are the real and imaginary parts of the complex Fresnel integral [57]. v can be expressed as

$$v = h_v \sqrt{\frac{2}{\lambda}\left(\frac{1}{r_{v1}} + \frac{1}{r_{v2}}\right)} \tag{4.49}$$

Where h_v is the distance from the top of the obstacle to the line connecting the transmit and receive antennas, and r_{v1} and r_{v2} are the distances from the top of the obstacle to the transmit and receive antennas, respectively. To simplify the calculation of diffraction loss, J_{dB} can be approximated as [84]

$$J_{dB}(v)=\begin{cases} 6.9+20\log\left[\sqrt{(v-0.1)^2}+v-0.1\right], v\ge-0.78 \\ 0, v\le-0.78 \end{cases}$$

(4.50)

The last step is to model the path loss in NLOS scenarios. The proposed NLOS path loss model in this book consists of two distance factors. The first distance factor follows the traditional power-law form d^{-n}, given by $(d_1+d_2+L_{bus}+L_{car})^{-n_1}$, where n_1 is the equivalent path loss exponent. The second distance factor reflects the impact of large vehicle obstructions on diffraction loss. To account for the influence of the distance between the mobile station and the bus on diffraction loss, the second distance factor can be expressed as $(d_1d_2)^{-n_2}$. It should be noted that these distance factors are heuristic models, which aim to improve the fitting accuracy of the model while still adhering to the physical logic of wireless propagation.

In summary, the channel gain model that incorporates the effect of large vehicle obstructions in NLOS scenarios can be expressed as follows:

$$P_G = P_{G_1} + P_{G_2} = \frac{G_1}{(d)^{n_1}} + \frac{G_2}{(d_1d_2)^{n_2}}$$

(4.51)

Here, n_1 and n_2 represent the equivalent path loss exponent, while G_1 and G_2 denote the reference gain levels. The parameters of the V2V static measurement path loss model are obtained through LS regression fitting. Table 4.5 lists the estimated values of G_1, G_2, n_1, and n_2 for different scenarios. From Table 4.5, it can be observed that n_1 and n_2 fluctuate around 2 and 4, respectively, in various scenarios. Furthermore, when the transmitter is far away from the bus (i.e., links "NLOS-CC-Location1-2" and "NLOS-CC-Location2-2"), the estimated value of G_2 is smaller, indicating a less impact on the loss estimation. For links "NLOS-CC-Location1-1" and "NLOS-CC-Location1-2," the estimated value of G_2 is larger compared to the results of links "NLOS-CC-Location2-1" and "NLOS-CC-Location2-2." This suggests that when the bus is placed in an open area, the impact of diffraction on the propagation loss is more significant because there are no other scattering objects available to provide energy from reflection paths.

To further evaluate the accuracy of the proposed statistical model, the coefficient of determination R^2 is analyzed. Table 4.5 summarizes the estimated values of R^2 for the proposed path loss model. It can be observed that R^2 is consistently greater than 0.7, indicating that the proposed path loss model in this section can accurately predict the measurement results. For comparison purposes, Figures 4.21 and 4.22 depict the curves of the proposed path loss model. It is important to note that Figures 4.21 and 4.22 show the path loss, which is the reciprocal of the channel gain.

TABLE 4.5 Parameters of Path Loss Model for V2V Static Measurement Link

Static Measurement Link Name	G_1	G_2	n_1	n_2	R^2
LOS-CC	Two-ray model				0.74
LOS-BC-Location1	Single knife-edge diffraction				0.71
LOS-BC-Location2	model/Two-ray model				0.79
	Two-ray model				
NLOS-CC-Location1-1	1.22×10^{-6}	6.36×10^{-6}	2.08	4.02	0.85
NLOS-CC-Location1-2	1.10×10^{-6}	2.02×10^{-6}	1.90	3.97	0.77
NLOS-CC-Location2-1	1.00×10^{-6}	4.76×10^{-6}	2.18	3.82	0.88
NLOS-CC-Location2-2	1.23×10^{-6}	1.17×10^{-6}	1.92	3.91	0.80

4.4.1.2 Shadow Fading

Shadow fading modeling is crucial for describing and quantifying the effects of large vehicle obstructions. By subtracting the distance-dependent median from the estimated path loss, the remaining component X_{LS} represents the shadow fading. In the logarithmic domain, X_{LS} is typically modeled as a zero-mean Gaussian distributed random variable with a standard deviation of σ_{LS}.

Figure 4.23 illustrates examples of CDF of shadow fading for the "LOS-CC" and "NLOS-CC-Location1-2" links, showing that a zero-mean Gaussian distribution accurately captures the variations of shadow fading in the logarithmic domain. Table 4.6 summarizes the measured standard deviation of shadow fading σ_{LS} for each link. It can be observed that in general, ls is smaller for LOS scenarios compared to NLOS scenarios. Additionally, the measured LS values are close to the results reported in the existing literature. For example, reference [82] reports σ_{LS} values of 1.7 dB for urban scenarios and 2.1 dB for suburban scenarios, reference [85] reports σ_{LS} of 3.4 dB, and reference [44] reports ls of 3.2 dB for rural scenarios.

Furthermore, the autocorrelation properties of shadow fading were investigated. The autocorrelation coefficient $\hat{\rho}_{auto}$ of shadow fading can be estimated using equation (4.14). Since the static test system operates in a distance-triggered mode, the time variable t in equation (4.14) is replaced with the distance d in the case of estimating $\hat{\rho}_{auto}$. This book adopts the classical exponential decay model [18] to describe the variation of $\hat{\rho}_{auto}$. Similar to previous discussions, the decorrelation distance $d_{cor,LS}$ of shadow fading is defined as the relative distance difference where the autocorrelation coefficient decreases to $\frac{1}{e}$.

The measurement results demonstrate that the exponential decay model effectively captures the characteristics of autocorrelation in static V2V scenarios. Table 4.6 summarizes the estimated values of $d_{cor,LS}$ for different scenarios, and the corresponding results align closely with reference [34]. Additionally, from the comparison between the "LOS-CC" and "LOS-BC-Location2" links, it can be observed that placing the transmitting antenna on top of the bus in LOS scenarios increases $d_{cor,LS}$. However, when the communication link is obstructed by the bus roof (i.e., "LOS-BC-Location1" link), $d_{cor,LS}$ decreases. In NLOS scenarios, the decorrelation distances for different links are relatively close to each other.

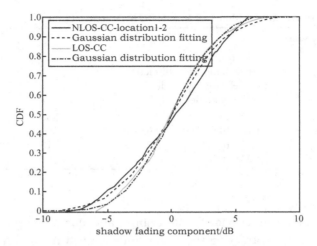

FIGURE 4.23   Example of CDF of shadow fading in V2V static measurements.

TABLE 4.6   Parameters of Shadow Fading Model for V2V
Static Measurement Link

Static Measurement Link Name	σ_{LS} / dB	$d_{cor,LS}$ / m
LOS-CC	2.34	8.84
LOS-BC-Location1	2.20	15.59
LOS-BC-Location2	2.62	22.07
NLOS-CC-Location1-1	2.86	9.32
NLOS-CC-Location1-2	3.50	13.79
NLOS-CC-Location2-1	2.79	13.28
NLOS-CC-Location2-2	3.30	14.18

4.4.1.3 Small-Scale Fading

To investigate the small-scale characteristics of V2V channels under large vehicle obstruction, this book excludes the local median (i.e., path loss and shadow fading) within each sliding window of 40 wavelengths. All space-time samples are then used for estimating the small-scale parameters. In static measurements, the 40-wavelength window ensures the WSS property of the local samples and provides a sufficient amount of data for extracting the small-scale parameters.

Firstly, the AIC method is employed to analyze the distribution characteristics of small-scale fading. Four commonly encountered fading distributions in V2V environments are selected as the objects of investigation [4,54]: Ricean, Rayleigh, Nakagami, and Weibull distributions. These four distributions are widely used for modeling fading channels in V2V environments. However, there is still a significant lack of comparative analysis to determine which distribution is most suitable for V2V channels. In this section, maximum likelihood estimation is performed on the parameters of these four distributions within a sliding window of 40 times the wavelength, and the AIC is used to compare the fitting characteristics of the distributions. Figure 4.24 presents an example of comparing the AIC

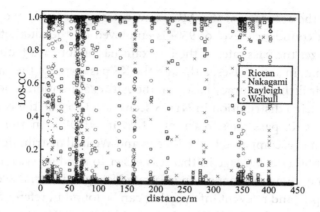

FIGURE 4.24   AIC Weight comparison of four alternative distributions for static V2V "LOS-CC" link.

weights of the four alternative distributions under the "LOS-CC" link. The distribution with the highest AIC weight exhibits the optimal fitting characteristics. Table 4.7 summarizes the statistical proportions of the optimal distributions for small-scale fading based on the AIC test, revealing the following observations.

1. The Nakagami distribution exhibits the optimal fitting characteristics in LOS scenarios, with a proportion exceeding 63%. Particularly, in the "LOS-CC" link, the proportion of optimal fitting exceeds 75%. This finding is consistent with the measurement results reported in reference [29]. In NLOS scenarios, the Nakagami distribution also demonstrates the optimal fitting characteristics in over 35.3% of the cases.

2. The Weibull distribution exhibits the optimal fitting characteristics in NLOS scenarios, with a proportion exceeding 45%. This finding is consistent with the conclusion drawn in reference [77]. However, the performance of the Weibull distribution is poor in LOS scenarios, with only 11.1% of cases demonstrating optimal fitting characteristics.

3. The Ricean distribution exhibits poor fitting performance in NLOS scenarios. In LOS scenarios, the Ricean distribution demonstrates good distribution fitting performance in the "LOS-BC-Location 1" and "LOS-BC-Location 2" links. However, in the "LOS-CC" link, the proportion of optimal fitting for the Ricean distribution is only 17%. This suggests that the Ricean distribution is only suitable for cases where the transmitting antenna is positioned vertically higher.

4. The Rayleigh distribution exhibits poor fitting performance in all static V2V links. This is because the Rayleigh distribution describes the worst-case fading scenario where there are no dominant paths and a large number of multipath components exist in the environment. However, in the NLOS scenarios measured in this book, there are still some strong multipath components present, such as diffracted and reflected paths. Therefore, the prerequisites of the Rayleigh distribution are not fully satisfied in this case.

To further ensure the accuracy of the AIC examination, a secondary examination using the KS test with a confidence level of 95% is employed on the alternative distributions. Table 4.7 summarizes the pass rates of the KS examination for each fading distribution. It can be observed that, in some cases, although the KS pass rates of the candidate distributions are high, their fitting accuracy based on the AIC examination is not high, as seen in the case of the Rayleigh distribution in NLOS scenarios. Overall, the Nakagami distribution has the highest KS pass rates, reaching 84.1% in NLOS scenarios and 77% in LOS scenarios. Furthermore, compared to the Ricean and Weibull distributions, the Nakagami distribution has another advantage in that it is easier to conduct system-level theoretical analysis. Research on how to generate an accurate Nakagami (correlated) fading channel is also more in-depth, and relevant information can be found in references [86–90]. Based on the above, in the modeling of fading channels in static V2V scenarios, the Nakagami distribution is adopted in this section.

Next, we analyze the parameter characteristics of the Nakagami distribution for fading channels under static large vehicle obstructions. The PDF of the received signal envelope r in the Nakagami fading channel can be expressed as follows:

$$p(r;m,\Omega)=\frac{2}{\mathrm{Gm}(m)}\left(\frac{m}{\Omega}\right)^2 r^{2m-1}\exp\left(-\frac{m}{\Omega}r^2\right) \tag{4.52}$$

Here, $\mathrm{Gm}(m)$ represents the gamma function, and Ω is the second moment of the distribution. The shape factor m in the Nakagami distribution is used to determine the severity of small-scale fading and can be obtained through maximum likelihood estimation. A smaller value of m indicates more severe fading. However, when $m=1$, the Nakagami distribution approximates the Rayleigh distribution.

Figure 4.25 presents the CDF statistics of the shape factor m in the Nakagami distribution for static V2V environments. The estimated values of m are much higher in LOS scenarios

TABLE 4.7 Optimal Distribution Proportions of Small-Scale Fading based on AIC and KS in the V2V Static Environment

Small-Scale Fading Distribution	Ricean		Nakagami	
Static Measurement Link Name	AIC	KS	AIC	KS
LOS-CC	17.49%	57.43%	75.21%	76.21%
LOS-BC-Location1	29.71%	66.57%	57.14%	80.75%
LOS-BC-Location2	30.32%	62.01%	55.08%	75.21%
LOS, All links	24.82%	61.23%	63.98%	77.00%
NLOS-CC-Location1-1	8.22%	71.18%	47.88%	83.69%
	30.51%	84.96%	40.36%	85.26%
NLOS-CC-Location1-2	20.94%	88.74%	20.85%	87.70%
	18.71%	78.40%	34.90%	80.94%
NLOS-CC-Location2-1	18.81%	81.93%	35.34%	84.10%
NLOS-CC-Location2-2				
NLOS, All links				

(Continued)

TABLE 4.7 *(Continued)* Optimal Distribution Proportions of Small-Scale Fading based on AIC and KS in the V2V Static Environment

Small-Scale Fading Distribution	Rayleigh		Weibull	
Static Measurement Link Name	AIC	KS	AIC	KS
LOS-CC	0.05%	3.83%	7.25%	38.03%
LOS-BC-Location1	0.11%	4.62%	13.04%	48.46%
LOS-BC-Location2	0.11%	4.34%	14.49%	50.35%
LOS, All links	0.08%	4.20%	11.12%	44.75%
NLOS-CC-Location1-1	0.47%	41.91%	43.42%	82.83%
NLOS-CC-Location1-2	0.58%	37.74%	28.54%	83.86%
NLOS-CC-Location2-1	1.53%	67.70%	56.67%	91.29%
NLOS-CC-Location2-2	0.55%	38.17%	45.84%	81.31%
NLOS, All links	0.79%	46.65%	45.06%	84.88%

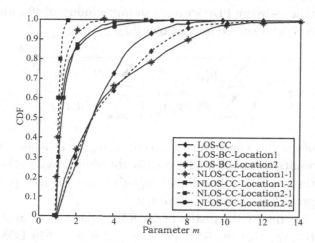

FIGURE 4.25 CDF of Nakagami distribution parameter m in static V2V environment.

compared to NLOS scenarios. When the transmitting antenna is positioned at a higher vertical location, such as in the "NLOS-BC-Location 1" and "NLOS-BC-Location 2" links, the value of m is larger. However, when the transmitting antenna is close to the bus, as in the "NLOS-CC-Location 1-1" and "NLOS-CC-Location 2-1" links, the value of m decreases due to strong shadowing effects. Table 4.8 summarizes the statistical characteristics of the shape factor m in the Nakagami distribution for V2V static environments. It is found that m approaches 1 in NLOS scenarios, indicating that the obstructions of the bus significantly intensify the small-scale fading of the wireless channel. In LOS scenarios, the V2V channels in urban areas exhibit fading characteristics that are "better than Rayleigh," with an average value of $m=3.45$. This result is also consistent with the measurement results reported in reference [29].

4.4.1.4 Delay Spread

The Root Mean Square (RMS) delay spread is widely used as a metric for time dispersion and frequency-selective channels [2] and is an important parameter for describing wideband wireless channels. The RMS delay spread can be defined using the Averaged Power

TABLE 4.8   Statistics of Nakagami Distribution Parameter m in the V2V Static Environment

Nakagami Distribution Parameter m	Mean	Standard Deviation	50%	90%
LOS-CC	3.17	1.65	2.84	5.33
LOS-BC-Location1	3.58	2.25	2.93	6.82
LOS-BC-Location2	3.76	2.72	2.84	7.90
LOS, All links	3.45	2.18	2.86	6.61
NLOS-CC-Location1-1	1.22	0.42	1.08	1.75
NLOS-CC-Location1-2	1.43	0.71	1.15	2.33
NLOS-CC-Location2-1	1.04	0.11	1.01	1.19
NLOS-CC-Location2-2	1.42	0.89	1.07	2.42
NLOS, All links	1.27	0.65	1.06	1.82

Delay Profile (APDP) $\overline{P}$. Unlike equation (4.13), the APDP in static measurements requires spatial averaging of the transient PDP within a sliding window of 40 times the wavelength. The definition of RMS delay spread is as follows [57].

$$\tau_{\text{rms}} = \sqrt{\frac{\sum_p \overline{P}(d,\tau_p)\tau_p^2}{\sum_p \overline{P}(d,\tau_p)} - \left(\frac{\sum_p \overline{P}(d,\tau_p)\tau_p}{\sum_p \overline{P}(d,\tau_p)}\right)^2} \tag{4.53}$$

In the calculation of the RMS delay spread, to eliminate the influence of low SNR signals on the estimation results, this book applies a noise threshold level that is 6 dB higher as the data processing threshold [91]. All multipath signals below the data processing threshold are excluded to improve the estimation accuracy.

Figure 4.26 presents the CDF statistics of the RMS delay spread in static V2V environments. It can be observed that the RMS delay spread is lowest in LOS scenarios, with a mean value of only 59 ns. In NLOS scenarios, the estimated values of the RMS delay spread exceed 500 ns at certain local positions. Here, it is worth noting that in most measurements of V2V LOS scenarios, the RMS delay spread is relatively small: references [85,91] indicate that the delay spread in V2V highway, rural, and urban environments is around 50 ns; reference [92] reports delay spreads of 47 and 75 ns in LOS scenarios at street crossings and tunnels, respectively. However, reference [93] points out that in urban LOS scenarios, a significant amount of backscattered waves may still exist, resulting in a delay spread of up to 500 ns. In NLOS scenarios, the measured RMS delay spreads tend to be larger in other references: reference [35] reports a delay spread of 173 ns in an NLOS scenario at a street crossing, which is 130 ns higher than the LOS scenario. Additionally, traffic congestion and large vehicle obstructions can cause delay spreads of 136 and 153 ns, respectively. However, the large vehicle obstruction phenomenon mentioned in reference [35] only occurs occasionally in dynamic V2V measurements. Therefore, the measurement results in this book can more effectively and accurately reflect the impact of large vehicle obstructions on the RMS delay spread. From Figure 4.26, it can be seen that when the mobile station is close to the bus, the RMS delay spread locally reaches 400 ns, a value significantly higher than that reported in the existing literature.

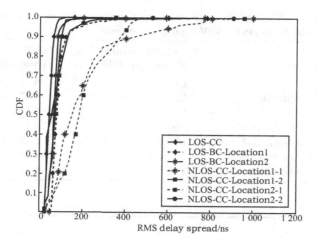

FIGURE 4.26 RMS delay spread in static V2V environment.

TABLE 4.9 Statistics of RMS Delay Spread in the V2V Static Environment

τ_{rms} / ns	Mean	Standard Deviation	50%	90%
LOS-CC	45.00	14.08	43.61	60.07
LOS-BC-Location1	87.34	47.62	75.39	122.7
LOS-BC-Location2	63.50	20.27	62.31	82.28
LOS, All links	59.83	30.37	55.35	84.05
NLOS-CC-Location1-1	200.2	165.7	148.3	435.6
NLOS-CC-Location1-2	67.95	22.85	66.72	94.94
NLOS-CC-Location2-1	201.1	106.6	182.3	375.7
NLOS-CC-Location2-2	68.84	51.06	52.80	126.4
NLOS, All links	134.6	121.9	93.1	283.5

Table 4.9 summarizes the estimated values of the RMS delay spread for each link. It can be observed that when $d_2 = 0.1\,\text{m}$, the measured RMS delay spread is often higher. This is due to the presence of strong shadowing effects in such cases.

4.4.1.5 Channel Parameter Correlations

To perform joint simulations of second-order statistical characteristics (i.e., shadow fading and delay spread) in V2V static environments, it is necessary to obtain the cross-correlation information among the second-order statistical parameters. If we exclude the influence of varying distances between the transmitter and receiver, both the RMS delay spread and shadow fading follow log-normal distributions, i.e., $10\log(\tau_{rms})$ and $10\log(X_{LS})$ follow normal distributions in the logarithmic domain. The estimates of the standard deviations for both parameters in the logarithmic dimension are summarized in Tables 4.6 and 4.10. From Table 4.10, it can be observed that the standard deviation of delay spread is generally lower in LOS scenarios, while smaller values of d_2 correspond to larger standard deviations of delay spread in NLOS scenarios.

References [94,95] based on measurements indicate the existence of correlation between channel shadow fading and delay spread. Figure 4.27 provides an example of the logarithmic

TABLE 4.10   Standard Deviation and Cross-Correlation Coefficient of Shadow
Fading/Delay Spread in the V2V Static Environment

τ_{rms} / ns	Standard Deviation of Delay Spread/dB	Cross-Correlation Coefficient
LOS-CC	1.14	−0.52
LOS-BC-Location1	1.39	−0.82
LOS-BC-Location2	1.24	−0.67
NLOS-CC-Location1-1	2.99	−0.21
NLOS-CC-Location1-2	1.25	−0.57
NLOS-CC-Location2-1	2.71	−0.19
NLOS-CC-Location2-2	2.51	−0.32

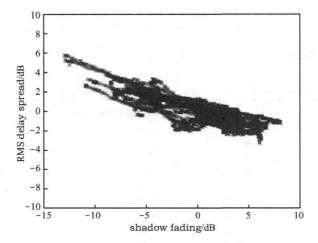

FIGURE 4.27   Example of cross-correlation characteristics of shadow fading and delay spread (with distance factor removed) in link "LOS-BC-Location1."

domain correlation between shadow fading and delay spread in the "LOS-BC-Location 1" link after removing the influence of the distance factor. It can be observed that there is a significant negative correlation between the two, which is consistent with the analytical analysis in reference [95]. Table 4.10 summarizes the cross-correlation coefficients between shadow fading and delay spread in the V2V static environment. Negative values of the cross-correlation coefficients indicate that larger delay spreads result in stronger shadow fading. Additionally, it was found in this book that the presence of large vehicle obstructions reduces the correlation between shadow fading and delay spread, especially when $d_2 = 0\,\text{m}$, where the absolute value of the correlation coefficient decreases to around 0.2.

4.4.2 Channel Modeling in Street Crossing Scenarios

This section presents the study and establishment of V2V wireless channel propagation models in suburban street crossing scenarios using V2V dynamic measurement data. First, the structural characteristics of two types of street crossing scenarios are described in detail. Then, the wireless channel parameters in each type of scenario are investigated in sequence.

4.4.2.1 Description and Discussion of Street Crossing Scenarios

The surrounding environmental characteristics of street crossing scenarios are similar to those described in Section 4.2. In this study, two specific street crossings were selected. The first street crossing is located in the Otaniemi area, east of Tapiola, and its aerial view is shown in Figure 4.28a. There are buildings and trees with a height of approximately 5 m on both sides of the road. The second street crossing is located on the west side of Tapiola, and its aerial view is shown in Figure 4.28b. There are a large number of trees on both sides of the road, and the road in the street crossing has a significant curvature, which leads to complete obstruction of LOS propagation when the distance between two vehicles is relatively long. In these two street crossings, measurements were conducted in three types of propagation scenarios.

Scenario 1 In this scenario, two communicating vehicles travel in the same direction before reaching the street crossing. After passing through the street crossing, the two vehicles separate and travel along different roads, as shown in Figure 4.28a. Prior to their separation, the receiving vehicle remains in front of the transmitting vehicle. In Scenario 1, after the vehicles separate at the street crossing, they are in different multipath environments, and the LOS propagation is completely obstructed by the presence of roadside buildings and trees. In Scenario 1, a total of 3,200 complete sets of MIMO complex channel spectral data were measured.

Scenario 2 In this scenario, two communicating vehicles travel in different directions and pass through the street crossing simultaneously, as shown in Figure 4.28b. Due to the curved shape of the street crossing, LOS propagation is completely obstructed by

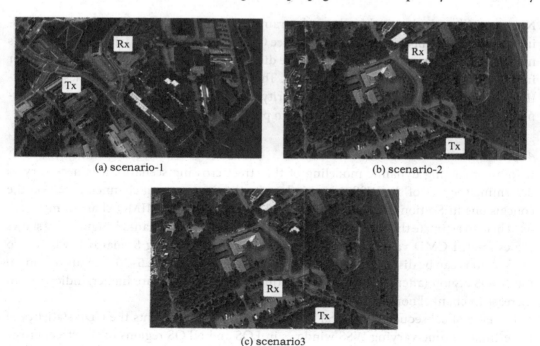

(a) scenario-1

(b) scenario-2

(c) scenario3

FIGURE 4.28 Top view of suburban street crossing scenario.

the roadside trees both before and after the street crossing. In this case, the received multipath signals mainly come from the backscattering of roadside obstacles. In Scenario 2, a total of 3,800 complete sets of MIMO complex channel spectral data were measured.

Scenario 3 The street crossing used in this scenario is the same as Scenario 2. In the measurement, the receiving vehicle is stationary at the pedestrian crosswalk of the street crossing, with the reference direction of the receiving antenna array indicated by the dashed arrow in 4.28c. The transmitting vehicle moves along the solid trajectory and passes by the receiving vehicle. Due to the curved shape of the road and the presence of roadside trees, LOS propagation is completely obstructed at the beginning and end of the entire moving trajectory. In Scenario 3, a total of 3,000 complete sets of MIMO complex channel spectral data were measured.

To model the wireless channel in street crossing scenarios more reasonably and effectively, the measurements at the street crossing are divided into the following two regions

- LOS region: this region refers to the area where there is LOS signal propagation between the two communicating vehicles, and they are in the same scattering environment.

- NLOS region: this region refers to the area where the LOS propagation is completely obstructed by buildings and trees due to the presence of the street crossing.

Next, we will study the wireless channel parameters in different regions of the street crossing scenario. It is important to note that since GPS positioning of the driving locations was not used in the dynamic measurements, the distance between the transmitter and receiver in the NLOS region cannot be determined. Therefore, in the subsequent studies, this section will only discuss channel non-stationarity, fast fading, delay, and angular spread phenomena, without analyzing the variations in path loss and shadow fading.

4.4.2.2 Wide-Sense Stationary Interval

Before conducting dynamic modeling of the street crossing scenario, it is necessary to determine the size of the WSS time window for the time-varying channel. Based on the conclusions in Section 4.3.1, the CMD based on the complete MIMO channel matrix is used here to estimate the WSS time window of the time-varying channel. Figure 4.29 shows the estimated CMD values at any given time in Street Crossing Scenario 1, where two WSS regions can be distinguished: the 0–5 seconds region and the 5–20 seconds region. In the NLOS region (after 20 seconds), the estimated CMD values are higher, indicating an increase in channel non-stationarity.

For ease of subsequent statistical analysis, Figure 4.30a presents the CDF statistics of the estimated time-varying WSS windows in LOS and NLOS regions of the street crossing scenario. Table 4.11 summarizes the statistical characteristics of the WSS windows in V2V street crossing scenarios, indicating that the WSS window in the NLOS region is

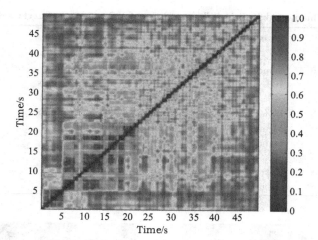

FIGURE 4.29 Example of CMD estimation in V2V street crossing Scenario 1.

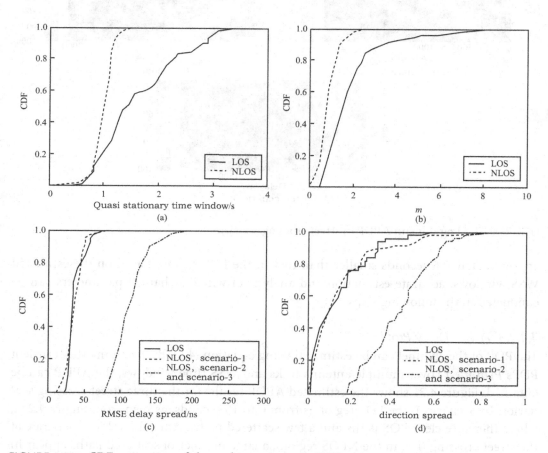

FIGURE 4.30 CDF estimation of channel parameters in V2V street crossing Scenario 1.

TABLE 4.11   WSS Window and Statistical Characteristics of m Parameter in Street Crossing Scenario 1

Parameters	Statistical Characteristics	LOS Region	NLOS Region
WSS Window	Mean	1.63 seconds	0.98 seconds
	Standard Deviation	0.72 seconds	0.17 seconds
	10%	0.82 seconds	0.76 seconds
	50%	1.43 seconds	0.99 seconds
	90%	2.83 seconds	1.17 seconds
m	Mean	1.86	0.84
	Standard Deviation	1.36	0.44
	10%	0.71	0.32
	50%	1.52	0.74
	90%	3.35	1.40

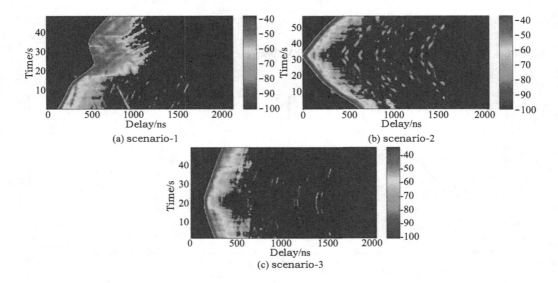

FIGURE 4.31   Log-domain APDP in street crossing scenario.

approximately 0.6 seconds smaller than that in the LOS region. Based on the estimated WSS windows, accurate estimation and analysis of wireless channel parameters can be conducted in the following steps.

4.4.2.3 Power Delay Profile

The PDP of the channel can be estimated using equation (4.5). By averaging the transient PDPs $P(m,n,t,\tau)$ over multiple antenna links and WSS time windows, the APDP can be obtained. Figure 4.31 shows the estimated APDP results for the three street crossing scenarios. In Scenario 1, the LOS region is from 0 to 15 seconds, as seen from Figure 4.24a, where there are clear LOS paths and a few scattered paths. After the vehicles separate at the street crossing (i.e., in the NLOS region), a large number of scattered paths appear in the APDP, accompanied by increased delay spread. In Scenarios 2 and 3, the LOS regions correspond to 25–45 seconds and 10–25 seconds, respectively. From Figure 4.24b and c, it can be observed that there are more scattered components at additional delays of 0–200 ns in the NLOS region. The APDP provides an overview of the channel propagation characteristics and will be used for the analysis of delay and angle dispersion phenomena in subsequent steps.

4.4.2.4 Small-Scale Fading

Similar to the previous discussion, in this case, the transient channel gains estimated using equation (4.6) are used. The small-scale fading distribution characteristics are analyzed within each WSS interval using $N_{TX} \times N_{RX} = 4 \times 30 = 120$ channels' MIMO test data, combined with the AIC. Before that, the influence of path loss and shadow fading is removed by subtracting the mean signal value within each WSS interval. In the AIC analysis, four common small-scale fading distributions for V2V channels are considered: Ricean, Rayleigh, Nakagami, and Weibull. Table 4.12 summarizes the proportions of the optimal distribution for small-scale fading based on AIC and the KS at a 95% confidence level in V2V street crossing scenarios. It can be observed that the Nakagami distribution exhibits the best fitting characteristics. It is important to note that, in order to use a unified metric for describing the small-scale fading channel, the distinction between LOS and NLOS regions and different street crossing scenarios is not considered in the determination of small-scale distribution characteristics in this section. Finally, the Nakagami distribution is adopted for the fading channel modeling in V2V street crossing scenarios in this section.

Figure 4.32 illustrates the variation of the parameter m based on maximum likelihood estimation over time in V2V street crossing scenarios, and it can be observed that in the LOS regions of Scenario 1 and Scenario 3, the value m is relatively high. This is due to the presence of strong LOS components and fewer scattered paths in the environment (as shown in Figure 4.31). In the LOS region of Scenario 2, however, the value m is only around 1–2, similar to the NLOS region, which is because in the LOS region of Scenario 2, there are still numerous long-delay multipath components present due to the presence of roadside trees and road traffic (as shown in Figure 4.31).

4.4.2.5 Delay Spread

The estimation method for the parameter of RMS delay spread, τ_{rms}, as a measure of channel time dispersion in street crossing scenarios, can be found in Section 4.4.1. Figure 4.33 depicts the variation of RMS delay spread over time in street crossing scenarios. It can be observed that there is a significant difference in τ_{rms} between the LOS and NLOS regions of Scenario

TABLE 4.12 Optimal Distribution Proportions of Small-Scale Fading Based on AIC and KSin V2V Street Crossing Scenario

Small-Scale Fading	AIC	KS
Ricean	2.92%	80.75%
Nakagami	80.78%	87.58%
Rayleigh	0.38%	45.45%
Weibull	15.92%	83.35%

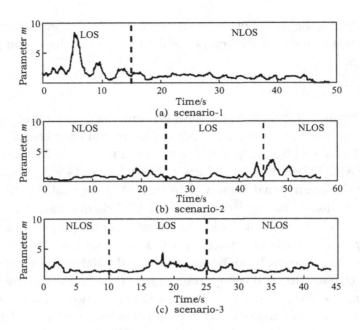

FIGURE 4.32   Variation of Nakagami distribution parameter *m* over time in V2V street crossing scenario.

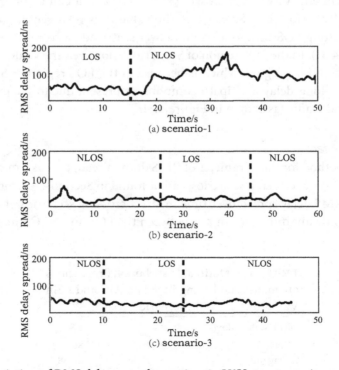

FIGURE 4.33   Variation of RMS delay spread over time in V2V street crossing scenario.

1. In the NLOS region, τ_{rms} is approximately 110 ns, which is nearly 70 ns higher than the LOS region. In Scenario 2 and Scenario 3, both the LOS and NLOS regions exhibit lower τ_{rms} values, and the results in the NLOS region are very close to the LOS region in Scenario 1. This is because in Scenario 2 and Scenario 3, the LOS link of the mobile station is completely blocked due to the curvature of the road and the presence of obstacles on both sides. Only a small number of multipath signals can reach the receiver through multiple reflections along the road (this characteristic is further validated in the subsequent analysis in the angular domain), which is consistent with the results reported in reference [96], where in the NLOS region, the delay spread is smaller than in the LOS region due to the scarcity of scattered waves. However, in the NLOS region of Scenario 1, the delay spread is significantly larger than in the LOS region. This is because in the NLOS region of Scenario 1, multipath signals can reach the receiver through diffraction and transmission along buildings and trees.

Table 4.13 summarizes the statistical characteristics τ_{rms} in street crossing scenario, and the CDF statistical results τ_{rms} are provided in Figure 4.30. In the aforementioned comparison, the NLOS region of Scenario 1 exhibits the highest delay spread, distinguishing it from the other scenarios. The estimated values τ_{rms} in the LOS regions of all three scenarios are similar to the results in the NLOS regions of Scenario 2 and Scenario 3.

4.4.2.6 Angular Domain Channel Characteristics

In order to further investigate the radio wave propagation mechanisms in street crossing scenario, the Bartlett beamforming method is used to analyze the directional channel parameters. The time-varying angular spectrum based on the Bartlett beamforming method can be expressed as [62]

$$\overline{P_{ang}}(\phi,\theta,t,\tau)=\alpha^{H}(\varphi,\theta)\cdot\tilde{R}(t,\tau)\cdot\alpha(\varphi,\theta) \tag{4.54}$$

Where φ and θ represent the azimuth and elevation angles of the multipath signals, respectively. Due to the limited number of antennas at the transmitter in the dynamic test system, the resolution of the departure angle estimation for multipath signals is restricted. Therefore, only the DOA of the multipath signals is considered. The matrix $\tilde{R}$ can be represented as

TABLE 4.13   Statistical Characteristics of Delay and Angle Spread in Street Crossing Scenario

Parameters	Statistical Characteristics	LOS Region	NLOS Region, Scenario 1	NLOS Region, Scenario 2 and 3
τ_{rms}	Mean	35.8 ns	111.5 ns	35.2 ns
	Standard Deviation	10.1 ns	22.2 ns	10.4 ns
	10%	26.3 ns	85.2 ns	23.7 ns
	50%	32.2 ns	109.1 ns	34.16 ns
	90%	53.8 ns	138.1 ns	47.3 ns
φ_{rms}	Mean	0.13	0.41	0.14
	Standard Deviation	0.13	0.14	0.15
	10%	0.01	0.11	0.01
	50%	0.09	0.41	0.09
	90%	0.32	0.61	0.32

$$\tilde{R}(t,\tau)=\frac{1}{(N_{TX})^2}\left(\sum_{n_T=1}^{N_{TX}}h(n_T,t,\tau)\right)\left(\sum_{n_T=1}^{N_{TX}}h(n_T,t,\tau)\right)^{H} \tag{4.55}$$

Where (n_T,t,τ) represents the channel impulse response matrix at the receiving end corresponding to the $n_{T\text{th}}$ transmit antenna at time t and delay τ. The steering vector $\alpha(\varphi,\theta)$ is defined as the projection of the antenna array manifold onto different polarization directions of the incident signal [97] and can be represented as

$$\alpha(\varphi,\theta)=A(\varphi,\theta)\cdot\hat{p} \tag{4.56}$$

Where $A(\varphi,\theta)$ represents the antenna's radiation pattern with specific polarization characteristics in a particular direction, and $\hat{p}$ is the polarization characteristic vector of the incident wave. To extract multipath signals effectively, this book only considers the case of the same polarization. By identifying the peaks of the angle spectrum based on Bartlett beamforming, the DOA information of the multipath signals can be obtained. In the multipath signal extraction mentioned above, a data processing threshold is set at 20 dB below the strongest signal to ensure that the extracted signals are valid multipath components.

Figures 4.34 and 4.35 show the estimated results of DOA azimuth and elevation angles at different times in the LOS and NLOS regions of the V2V crossroad scenario.

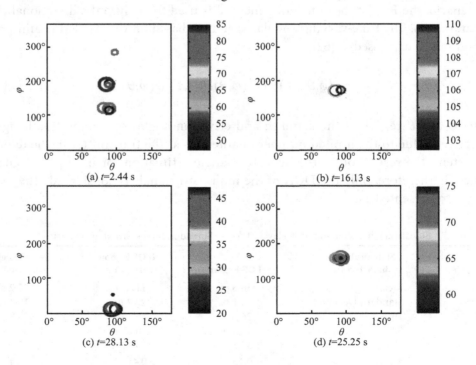

FIGURE 4.34 Example of DOA Azimuth and elevation estimation at different time instances in LOS region. ((a) and (b) are from Scenario 1; (c) is from Scenario 2; (d) is from Scenario 3. The color bar represents the propagation path length of multipath signals in meters, and the size of circular markers represents the strength of multipath signals.)

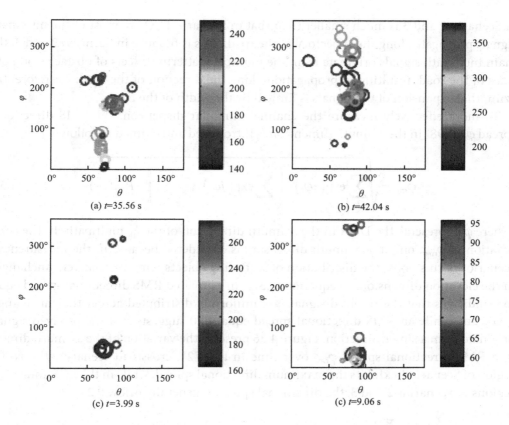

FIGURE 4.35 Example of DOA azimuth and elevation estimation at different time instances in NLOS region. ((a) and (b) are from Scenario 1; (c) is from Scenario 2; (d) is from Scenario 3. The color bar represents the propagation path length of multipath signals in meters, and the size of circular markers represents the strength of multipath signals.)

In this case, the angle spectrum is amplitude normalized within each WSS time window, and the propagation path length (i.e., the product of signal additional delay and the speed of light) is used to represent the temporal variation of the multipath signals. From Figure 4.34b–d, it can be observed that the angular dispersion phenomenon is relatively weak in the LOS region, and most of the multipath signals propagate approximately along the trajectory of the LOS signal. However, in Figure 4.34a, the multipath signals at the azimuth angles of 100° and 290° exhibit a certain degree of angular spreading, indicating that there could still be significant angular expansion in the LOS region due to the presence of numerous scattering objects on both sides of the road in the V2V environment. From Figures 4.35a and b, it can be seen that the NLOS region of Scenario 1 exhibits significant angular expansion. This is because when the LOS is completely obstructed by buildings and trees, the multipath signals in the NLOS region of Scenario 1 can only reach the receiver through diffraction over the top of buildings/trees and transmission through gaps between buildings/trees. Both propagation paths lead to significant angular spreading. Furthermore, Figures 4.35c and d show that the angular expansion in the NLOS regions of Scenario 2 and 3 is much smaller than that in Scenario 1, where most of the multipath signals propagate along the trajectory of the road. This is because in Scenario 2 and 3, the main multipath signals originate from the multiple scattering effects of obstacles on both sides of the road, resulting in propagation along the direction of the road. Therefore, the azimuthal expansion of the signals is limited by the width of the road.

To more effectively measure the channel's angular dispersion, the RMS directional spread φ_{rms} [98] in the azimuth dimension is introduced and defined as follows:

$$\varphi_{rms} = \sqrt{\sum_l \left| \exp(j\varphi_l) - \left(\sum_l \exp(j\varphi_l)\overline{P_{ang}}(\varphi_l) \right) \right|^2 \cdot \overline{P_{ang}}(\varphi_l)} \qquad (4.57)$$

Where φ_l represents the DOA in the azimuth dimension of the l_{th} multipath. In the computation of φ_{rms}, only the azimuth dimension is considered because in the measurement scenarios of this book, the distribution of scattering objects remains relatively unchanged in the elevation dimension. In equation (4.57), an estimated RMS directional spread equal to 1 indicates that the received signals are uniformly distributed across the entire angle spectrum, while an RMS directional spread equal to 0 suggests that the received signals originate from a single direction. Figure 4.36 presents the variation of the azimuth dimension RMS directional spread φ_{rms} over time in the V2V crossroad scenario. The NLOS region of Scenario 1 exhibits the maximum directional spread, while in the LOS and NLOS regions of Scenarios 2 and 3, the directional spread is generally below 0.2.

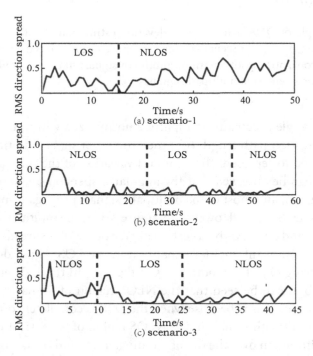

FIGURE 4.36 Variation of RMS directional spread in Azimuth dimension over time in V2V intersection scenario.

Figure 4.30d presents the CDF statistical results of the estimated φ_{rms} in the LOS and NLOS regions, with a separate column for the NLOS region of Scenario 1. From the graph, it can be observed that the NLOS region of Scenario 1 exhibits the maximum directional spread, while the NLOS regions of Scenarios 2 and 3 show φ_{rms} estimates that are very close to the estimates in the LOS regions of all three Scenarios. This observation aligns well with the previous analysis, which highlighted that the directional spread in Scenarios 2 and 3 is constrained by the width of the street. Table 4.13 summarizes the statistical characteristics of φ_{rms} in different Scenario categories. A comparison with existing measurement results from Reference [40], conducted in urban crossroad environments, reveals that the RMS directional spread ranges from 0.4 to 0.9, which is higher than the measurement results presented in this book. This difference is attributed to the richer presence of scattering objects in urban crossroad environments.

4.4.2.7 Channel Parameters Cross-Correlation

Extensive research [94,99,100] has shown that there exists a correlation between the delay spread and the angular spread of a channel. Therefore, a similar approach as in the previous analysis is adopted here to study the cross-correlation between these two parameters. First, the effects of time-domain factors on the delay spread $10\log(\tau_{rms})$ and angular spread $10\log(\varphi_{rms})$ are removed separately in the LOS and NLOS regions. Then, the logarithmic domain delay spread and angular spread follow Gaussian distributions with zero mean. Finally, the cross-correlation coefficient between them is estimated. Table 4.14 summarizes the cross-correlation coefficients between the delay spread and angular spread in different scenario categories, revealing the following observations.

1. The absolute values of the cross-correlation coefficients range between 0.1 and 0.5, indicating a correlation between the mechanisms causing time and angular dispersion. In contrast to the measurement results in urban and suburban areas reported in reference [94], negative correlation coefficients are observed in the V2V environment. This is attributed to the presence of numerous distant scatterers and the trench-like structure formed by abundant trees and buildings on both sides of the road, which typically induce smaller angular spreads.

2. The absolute values of the measured correlation coefficients are smaller than those reported in non-V2V scenarios [94,99,100]. In the aforementioned literature, the

TABLE 4.14   Cross-correlation Coefficient in Street Crossing Scenario

Propagation Region	Scenario	Cross-Correlation Coefficient
LOS Region	Scenario 1	−0.30
	Scenario 2	−0.13
	Scenario 3	−0.08
NLOS Region	Scenario 1	0.32
	Scenario 2	0.47
	Scenario 3	−0.14

absolute values of the correlation coefficients are often greater than 0.5. This can be attributed to the dynamic scatterers present in the V2V environment, which reduce the correlation between delay spread and angular spread.

3. No clear association is observed between the magnitude and sign of the correlation coefficients and the scenario types or LOS/NLOS regions in the measurement results. This is due to the stochastic nature of the distribution of various types of scatterers in the time-varying V2V environment, where different propagation mechanisms causing channel correlations can occur.

REFERENCES

[1] (He Ruisi) 何睿斯. 车载网络复杂场景下无线信道测量与建模研究. 北京: 北京交通大学, 2015.

[2] Bello P. Characterization of randomly time-variant linear channels. *IEEE Transactions on Communications Systems*, 1963, 11(4): 360–393.

[3] Yan L, He R, Lin S, et al. Cluster-based non-stationary channel modeling for vehicle-to-vehicle communications. *IEEE Antennas and Wireless Propagation Letters*, 2017, 16(1): 408–411.

[4] Molisch A F, Tufvesson F, Karedal J, et al. A survey on vehicle-to-vehicle propagation channels. *IEEE Wireless Communications*, 2009, 16(6): 12–22.

[5] Matz G. On non-WSSUS wireless fading channels. *IEEE Transactions on Wireless Communications*, 2005, 4(5): 2465–2478.

[6] Bultitude R, Charalambous C, Herben M, Brussaard G. Development of a model for realistic portrayal of random time variations on mobile radio channels. *Proceedings of URSI General Assembly, August 17–24, 2002, Maastricht, the Netherlands, USA*, 2002: 2119–2122.

[7] Matz G. Characterization of non-WSSUS fading dispersive channels. *IEEE ICC, May 11–15, 2003, Anchorage, USA*. Piscataway, NJ: IEEE Press, 2003: 2480–2484.

[8] Herdin M. *Non-Stationary Indoor MIMO Radio Channels*. Austria: Technische Universit¨at Wien, 2004: 37–47.

[9] Herdin M, Bonek E. A MIMO correlation matrix based metric for characterizing non-stationarity. *IST Mobile and Wireless Communications Summit, June 27–30, 2004, Lyon, France*. USA: IST, 2004: 1–5.

[10] Herdin M, Czink N, Ozcelik H, Bonek E. Correlation matrix distance, a meaningful measure for evaluation of non-stationary MIMO channels. *IEEE VTC, May 30–June 1, 2005, Stockholm, Sweden*. Piscataway, NJ: IEEE Press, 2005: 136–140.

[11] Renaudin O, Kolmonen V, Vainikainen P, Oestges C. Car-to-car channel models based on wideband MIMO measurements at 5.3 GHz. *Proceedings of IEEE EuCAP, March 23–27, 2009, Berlin, Germany*. Piscataway, NJ: IEEE Press, 2009: 635–639.

[12] Bernad O L. *Non-Stationarity in Vehicular Wireless Channels*. Austria: Technische Universitat Wien, 2012: 53–68.

[13] Renaudin O, Kolmonen V M, Vainikainen P, et al. Non-stationary narrowband MIMO inter-vehicle channel characterization in the 5-GHz band. *IEEE Transactions on Vehicular Technology*, 2010, 59(4): 2007–2015.

[14] Paier A, Zemen T, Bernad O L, Matz G, Karedal J, Czink N, Dumard C, Tufvesson F. Non-WSSUS vehicular channel characterization in highway and urban scenarios at 5.2 GHz using the local scattering function. *IEEE ITG WSA, February 26–27, 2008, Vienna, Austria*. Piscataway, NJ: IEEE Press, 2008: 9–15.

[15] Georgiou T. Distances and Riemannian metrics for spectral density functions. *IEEE Transactions on Signal Processing*, 2007, 55(8): 3995–4003.

[16] Bernad O L, Zemen T, Paier A, Matz G, Karedal J, Czink N, Dumard C, Tufvesson F, Hagenauer M, Molisch A F, Mecklenbrauker C F. Non-WSSUS vehicular channel characterization at 5.2 GHz-spectral divergence and time-variant coherence parameters. *URSI General Assembly, August 9–16, 2008, Chicago, USA.* USA: URSI, 2008: 9–16.

[17] Bernad O L, Zemen T, Paier A, Karedal J. Complexity reduction for vehicular channel estimation using the filter divergence measure. *IEEE ASILOMAR, November 7–10, 2010, Pacific Grove, Canada.* Piscataway, NJ: IEEE Press, 2010: 141–145.

[18] Gudmundson M. Correlation model for shadow fading in mobile radio systems. *Electronics Letters,* 1991, 27(23): 2145–2146.

[19] Gehring A, Steinbauer M, Gaspard I, Grigat M. Empirical channel stationarity in urban environments. *IET EPMCC, February 20–22, 2001, Vienna, Austria.* UK: IET, 2001: 1–6.

[20] Jamsa T, Kyosti P, Iinatti J. Correlation error metrics of simulated MIMO channels. *IEEE VTC, April 22–25, 2007, Dublin, Ireland.* Piscataway, NJ: IEEE Press, 2007: 407–412.

[21] Bultitude R, Brussaard G, Herben M. Radio channel modelling for terrestrial vehicular mobile applications. *IEEE MCAP, April 9–14, 2000, Davos, Switzerland.* Piscataway: IEEE Press, 2000: 1–5.

[22] Umansky D, Patzold M. Stationarity test for wireless communication channels. *IEEE GLOBECOM, November 30–December 4, 2009, Honolulu, USA.* Piscataway, NJ: IEEE Press, 2009: 1–6.

[23] Maurer J, Fugen T, Schafer T, Wiesbeck W. A new inter-vehicle communications (IVC) channel model. *IEEE VTC, September 26–29, 2004, Los Angeles, USA.* Piscataway, NJ: IEEE Press, 2004: 9–13.

[24] Biddlestone S, Redmill K, Miucic R, et al. An integrated 802.11p WAVE DSRC and vehicle traffic simulator with experimentally validated urban (LOS and NLOS) propagation models. *IEEE Transactions on Intelligent Transportation Systems,* 2012, 13(4): 1792–1802.

[25] Hosseini S A, Fleury M, Qadri N, et al. Improving propagation modeling in urban environments for vehicular Ad Hoc networks. *IEEE Transactions on Intelligent Transportation Systems,* 2011, 12(3): 705–716.

[26] Boban M, Vinhoza T, Ferreira M, et al. Impact of vehicles as obstacles in vehicular Ad Hoc networks. *IEEE Journal on Selected Areas in Communications,* 2011, 29(1): 15–28.

[27] Paier A, Karedal J, Czink N, Hofstetter H, Dumard C, Zemen T, Tufvesson F, Andreas F. Car-to-car radio channel measurements at 5 GHz: Pathloss, power-delay profile, and delay-doppler spectrum. *IEEE ISWCS, October 17–19, 2007, Trondheim, Norway.* Piscataway, NJ: IEEE Press, 2007: 224–228.

[28] Paier A, Karedal J, Czink N, Hofstetter H, Dumard C, Zemen T, Tufvesson F, Christoph. First results from car-to-car and car-to-infrastructure radio channel measurements at 5.2 GHz. *IEEE PIMRC, September 3–7, 2007, Athens, Greece.* Piscataway, NJ: IEEE Press, 2007: 1–5.

[29] Cheng L, Henty B E, Stancil D, et al. Mobile vehicle-to-vehicle narrow-band channel measurement and characterization of the 5.9 GHz dedicated short range communication (DSRC) frequency band. *IEEE Journal on Selected Areas in Communications,* 2007, 25(8): 1501–1516.

[30] Karedal J, Tufvesson F, Czink N, et al. A geometry-based stochastic MIMO model for vehicle-to-vehicle communications. *IEEE Transactions on Wireless Communications,* 2009, 8(7): 3646–3657.

[31] Cheng L, Henty B E, Cooper R, et al. A measurement study of time-scaled 802.11a waveforms over the mobile-to-mobile vehicular channel at 5.9 GHz. *IEEE Communications Magazine,* 2008, 46(5): 84–91.

[32] Rustako Jr A, Gans M, Owens G, et al. Attenuation and diffraction effects from truck blockage of an 11-GHz line-of-sight microcellular mobile radio path. *IEEE Transactions on Vehicular Technology,* 1991, 40(1): 211–215.

[33] Yamamoto A, Ogawa K, Horimatsu T, et al. Path-loss prediction models for intervehicle communication at 60 GHz. *IEEE Transactions on Vehicular Technology*, 2008, 57(1): 65–78.

[34] Abbas T, Katrin S, Karedal J, Tufvesson F. A measurement based shadow fading model for vehicle-to-vehicle network simulations. *International Journal of Antennas & Propagation*, 2015: 190607, 12 DOI: 10.1155/2015/190607.

[35] Bernad O L, Zemen T, Tufvesson F, et al. Delay and doppler spreads of non-stationary vehicular channels for safety relevant scenarios. *IEEE Transactions on Vehicular Technology*, 2014, 63(1): 82–93.

[36] Meireles R, Boban M, Steenkiste P, Tonguz O, Barros J. Experimental study on the impact of vehicular obstructions in VANETs. *IEEE VNC, December 13–15, 2010, Jersey City, USA.* Piscataway, NJ: IEEE Press, 2010: 338–345.

[37] Boban M, Meireles R, Barros J, Tonguz O, Steenkiste P. Exploiting the height of vehicles in vehicular communication. *IEEE VTC, November 14–16, 2011, Amsterdam, Netherlands.* Piscataway, NJ: IEEE Press, 2011: 163–170.

[38] Karedal J, Tufvesson F, Abbas T, Klemp O, Paier A, Bernadó L, Molisch A F. Radio channel measurements at street intersections for vehicle-to-vehicle safety applications. *IEEE VTC, May 16–19, 2010, Taipei, Taiwan, China.* Piscataway, NJ: IEEE Press, 2010: 1–5.

[39] Schack M, Nuckelt J, Geise R, Thiele L, Kürner T. Comparison of path loss measurements and predictions at urban crossroads for C2C communications. *IEEE EUCAP, April 11–15, 2011, Rome, Italy.* Piscataway, NJ: IEEE Press, 2011: 2896–2900.

[40] Abbas T, Karedal J, Tufvesson F, Klemp O, Paier A, Bernadó L, Molisch A F. Directional analysis of vehicle-to-vehicle propagation channels. *IEEE VTC, May 15–18, 2011, Yokohama, Japan.* Piscataway, NJ: IEEE Press, 2011: 1–5.

[41] Schumacher H, Tchouankem H, Nuckelt J, Kürner T, Zinchenko T, Leschke A, Wolf L. Vehicle-to-vehicle IEEE 802.11p performance measurements at urban intersections. *IEEE ICC, June 10–15, 2012, Ottawa, Canada.* Piscataway, NJ: IEEE Press, 2012: 7131–7135.

[42] Mangel T, Klemp O, Hartenstein H. 5.9 GHz inter-vehicle communication at intersections: A validated non-line-of-sight path-loss and fading model. *EURASIP Journal on Wireless Communications and Networking*, 2011, 2011(1): 1–11.

[43] Wang C X, Cheng X, Laurenson D I. Vehicle-to-vehicle channel modeling and measurements: Recent advances and future challenges. *IEEE Communications Magazine*, 2009, 47(11): 96–103.

[44] Sen I, Matolak D W. Vehicle-vehicle channel models for the 5-GHz band. *IEEE Transactions on Intelligent Transportation Systems*, 2008, 9(2): 235–245.

[45] Renaudin O, Kolmonen V M, Vainikainen P, et al. Wideband measurement-based modeling of inter-vehicle channels in the 5-GHz band. *IEEE Transactions on Vehicular Technology*, 2013, 62(8): 3531–3540.

[46] Chen Y, Dubey V K. Dynamic simulation model of indoor wideband directional channels. *IEEE Transactions on Vehicular Technology*, 2006, 55(2): 417–430.

[47] Chong C, Tan C M, Laurenson D I, et al. A novel wideband dynamic directional indoor channel model based on a Markov process. *IEEE Transactions on Wireless Communications*, 2005, 4(4): 1539–1552.

[48] Zwick T, Fischer C, Wiesbeck W. A stochastic multipath channel model including path directions for indoor environments. *IEEE Journal on Selected Areas in Communications*, 2002, 20(6): 1178–1192.

[49] Zwick T, Fischer C, Didascalou D, et al. A stochastic spatial channel model based on wave-propagation modeling. *IEEE Journal on Selected Areas in Communications*, 2000, 18(1): 6–15.

[50] Nielsen J Ø, Afanassiev V, Andersen J B. A dynamic model of the indoor channel. *Wireless Personal Communications*, 2001, 19(2): 91–120.

[51] Piechocki R J, McGeehan J P, Tsoulos G V. A new stochastic spatio-temporal propagation model (SSTPM) for mobile communications with antenna arrays. *IEEE Transactions on Communications*, 2001, 49(5): 855–862.

[52] Wang W, Jost T, Fiebig U, Koch W. Time-variant channel modeling with application to mobile radio based positioning. *IEEE GLOBECOM, December 3–7, 2012, Anaheim, USA.* Piscataway, NJ: IEEE Press, 2012: 5038–5043.

[53] Amiri K, Sun Y, Murphy P, Hunter C, Cavallaro J R, Sabharwal A. Warp, a unified wireless network testbed for education and research. *IEEE ICMSE, June 3–4, 2007, San Diego, USA.* Piscataway, NJ: IEEE Press, 2007: 53–54.

[54] Mecklenbrauker C F, Molisch A F, Karedal J, et al. Vehicular channel characterization and its implications for wireless system design and performance. *Proceedings of the IEEE*, 2011, 99(7): 1189–1212.

[55] Chu D. Polyphase codes with good periodic correlation properties. *IEEE Transactions on Information Theory*, 1972, 18(4): 531–532.

[56] Kolmonen V, Kivinen J, Vuokko L, et al. 5.3GHz MIMO radio channel sounder. *IEEE Transactions on Instrumentation and Measurement*, 2006, 55(4): 1263–1269.

[57] Molisch A F. *Wireless Communications*. 2nd ed. USA: Wiley, 2010: 102–235.

[58] Renaudin O. *Experimental Channel Characterization for Vehicle-to-Vehicle Communication Systems*. Belgium: Université Catholique de Louvain, 2013: 65–75.

[59] Karedal J, Wyne S, Almers P, et al. A measurement-based statistical model for industrial ultra-wideband channels. *IEEE Transactions on Wireless Communications*, 2007, 6(8): 3028–3037.

[60] He R, Renaudin O, Kolmonen V M, Haneda K, Zhong Z, Ai B, Oestges C. Non-stationarity characterization for vehicle-to-vehicle channels using correlation matrix distance and shadow fading correlation. *Progress in Electromagnetics Research Symposium, August 25–28, 2014, Guangzhou, China*, 2014: 2144–2148.

[61] Stoica P, Moses R L. *Spectral Analysis of Signals*. USA: Pearson/Prentice Hall Upper Saddle River, 2005: 319–322.

[62] Krim H, Viberg M. Two decades of array signal processing research: The parametric approach. *IEEE Signal Processing Magazine*, 1996, 13(4): 67–94.

[63] Czink N, Yin X, Ozcelik H, et al. Cluster characteristics in a MIMO indoor propagation environment. *IEEE Transactions on Wireless Communications*, 2007, 6(4): 1465–1475.

[64] Chen C, Jensen M A. A stochastic model of the time-variant MIMO channel based on experimental observations. *IEEE Transactions on Vehicular Technology*, 2009, 58(6): 2618–2625.

[65] Czink N, Herdin M, Ozcelik H, et al. Number of multipath clusters in indoor MIMO propagation environments. *Electronics Letters*, 2004, 40(23): 1498–1499.

[66] Steinbauer M, Ozcelik H, Hofstetter H, et al. How to quantify multipath separation. *IEICE Transactions on Electronics*, 2002, 85(3): 552–557.

[67] Czink N, Cera P, Salo J, Bonek E, Nuutinen J P, Ylitalo J. Automatic clustering of MIMO channel parameters using the multi-path component distance measure. *IEEE WPMC, September 18–22, Aalborg, Denmark.* Piscataway, NJ: IEEE Press, 2005: 1–5.

[68] Czink N, Cera P, Salo J, Bonek E, Nuutinen J P, Ylitalo J. A framework for automatic clustering of parametric MIMO channel data including path powers. *IEEE VTC, September 25–28, 2006, Montreal, Canada.* Piscataway, NJ: IEEE Press, 2006: 1–5.

[69] Czink N, Mecklenbrauker C. A novel automatic cluster tracking algorithm. *IEEE PIMRC, September 11–14, 2006, Helsinki, Finland.* Piscataway, NJ: IEEE Press, 2006: 1–5.

[70] Czink N. *The Random-Cluster Model: A Stochastic MIMO Channel Model for Broadband Wireless Communication Systems of the 3rd Generation and Beyond*. Austria: Technische Universit"at Wien, 2007: 99–123.

[71] Spencer Q H, Jeffs B D, Jensen M A, et al. Modeling the statistical time and angle of arrival characteristics of an indoor multipath channel. *IEEE Journal on Selected Areas in Communications*, 2000, 18(3): 347–360.

[72] Heddergott R, Bernhard U P, Fleury B H. Stochastic radio channel model for advanced indoor mobile communication systems. *IEEE PIMRC, September 1–4, 1997, Helsinki, Finland.* Volume 1. Piscataway, NJ: IEEE Press, 1997: 140–144.

[73] Spencer Q, Rice M, Jeffs B, Jensen M. A statistical model for angle of arrival in indoor multipath propagation. *IEEE VTC, May 4–7, 1997, Phoenix, USA.* Volume 3. Piscataway, NJ: IEEE Press, 1997: 1415–1419.

[74] Chong C, Tan C M, Laurenson D I, et al. A new statistical wideband spatio-temporal channel model for 5-GHz band WLAN systems. *IEEE Journal on Selected Areas in Communications,* 2003, 21(2): 139–150.

[75] Abdi A, Barger J A, Kaveh M. A parametric model for the distribution of the angle of arrival and the associated correlation function and power spectrum at the mobile station. *IEEE Transactions on Vehicular Technology,* 2002, 51(3): 425–434.

[76] Abdi A, Kaveh M. A space-time correlation model for multielement antenna systems in mobile fading channels. *IEEE Journal on Selected Areas in Communications,* 2002, 20(3): 550–560.

[77] Matolak D W. Channel modeling for vehicle-to-vehicle communications. *IEEE Communications Magazine,* 2008, 46(5): 76–83.

[78] Cassioli D, Win M Z, Molisch A F. The ultra-wide bandwidth indoor channel: From statistical model to simulations. *IEEE Journal on Selected Areas in Communications,* 2002, 20(6): 1247–1257.

[79] Lee W C. Estimate of local average power of a mobile radio signal. *IEEE Transactions on Vehicular Technology,* 1985, 34(1): 22–27.

[80] He R, Molisch A F, Tufvesson F, et al. Vehicle-to-vehicle channel models with large vehicle obstructions. *IEEE International Conference on Communications (ICC), June 10–14, 2014, Australia,* 2014: 5647–5652.

[81] Karedal J, Czink N, Paier A, et al. Path loss modeling for vehicle-to-vehicle communications. *IEEE Transactions on Vehicular Technology,* 2011, 60(1): 323–328.

[82] Rappaport T S. *Wireless Communications: Principles and Practice.* USA: Prentice Hall, 1996: 69–138.

[83] ITU-R. *Propagation by Diffraction: P.526-12. 02-2012.* Geneva: International Telecommunication Union Radiocommunication Sector, 2012: 15–32.

[84] Kunisch J, Pamp J. Wideband car-to-car radio channel measurements and model at 5.9 GHz. *IEEE VTC, September 21–24, 2008, Calgary, Canada.* Piscataway, NJ: IEEE Press, 2008: 1–5.

[85] Laourine A, Alouini M S, Affes S, et al. On the performance analysis of composite multipath/shadowing channels using the G-distribution. *IEEE Transactions on Communications,* 2009, 57(4): 1162–1170.

[86] Filho J, Yacoub M, Fraidenraich G. A simple accurate method for generating autocorrelated nakagami-m envelope sequences. *IEEE Communications Letters,* 2007, 11(3): 231–233.

[87] Karagiannidis G K, Zogas D A, Kotsopoulos S A. On the multivariate nakagami-m distribution with exponential correlation. *IEEE Transactions on Communications,* 2003, 51(8): 1240–1244.

[88] Zhang Q. A decomposition technique for efficient generation of correlated nakagami fading channels. *IEEE Journal on Selected Areas in Communications,* 2000, 18(11): 2385–2392.

[89] Chen Y, Tellambura C. Distribution functions of selection combiner output in equally correlated Rayleigh, Rician, and Nakagami-m fading channels. *IEEE Transactions on Communications,* 2004, 52(11): 1948–1956.

[90] Molisch A F, Steinbauer M. Condensed parameters for characterizing wideband mobile radio channels. *International Journal of Wireless Information Networks,* 1999, 6(3): 133–154.

[91] Cheng L, Henty B, Cooper R, Stancil D D, Bai F. Multi-path propagation measurements for vehicular networks at 5.9 GHz. *IEEE WCNC, March 31–April 3, 2008, Las Vegas, USA.* Piscataway, NJ: IEEE Press, 2008: 1239–1244.

[92] Tan I, Tan G W, Laberteaux K, Bahai A. Measurement and analysis of wireless channel impairments in DSRC vehicular communications. *IEEE ICC, May 19-23, 2008, Beijing, China*. Piscataway, NJ: IEEE Press, 2008: 4882–4888.

[93] Algans A, Pedersen K I, Mogensen P E. Experimental analysis of the joint statistical properties of azimuth spread, delay spread, and shadow fading. *IEEE Journal on Selected Areas in Communications*, 2002, 20(3): 523–531.

[94] Greenstein L J, Erceg V, Yeh Y S, et al. A new path-gain/delay-spread propagation model for digital cellular channels. *IEEE Transactions on Vehicular Technology*, 1997, 46(2): 477–485.

[95] Abbas T, Bernado L, Thiel A, et al. Radio channel properties for vehicular communication: Merging lanes versus urban intersections. *IEEE Vehicular Technology Magazine*, 2013, 8(4): 27–34.

[96] Sarkis R. *Antenna Arrays for Direction of Arrival Estimation and Imaging: From Mutual Coupling Analysis to Real-World Design*. Belgium: Universite Catholique de Louvain, 2011: 73–114.

[97] Fleury B H. First-and second-order characterization of direction dispersion and space selectivity in the radio channel. *IEEE Transactions on Information Theory*, 2000, 46(6): 2027–2044.

[98] Asplund H, Glazunov A A, Molisch A F, et al. The COST 259 directional channel model-part II: Macrocells. *IEEE Transactions on Wireless Communications*, 2006, 5(12): 3434–3450.

[99] Pedersen K I, Mogensen P E, Fleury B H. A stochastic model of the temporal and azimuthal dispersion seen at the base station in outdoor propagation environments. *IEEE Transactions on Vehicular Technology*, 2000, 49(2): 437–447.

[100] Cheng L, Henty B E, Bai F, Stancil D D. Highway and rural propagation channel modeling for vehicle-to-vehicle communications at 5.9 GHz. *IEEE AP-S, July 5-11, 2008, San Diego, USA*. Piscataway, NJ: IEEE Press, 2008: 1–4.

CHAPTER 5

Rail Transit Channel Measurement and Modeling

THE CHARACTERISTICS OF COMMUNICATION channel in the field of rail transit have attracted great attention in recent years. This chapter will introduce the classification standard of communication scenarios in high-speed railway environment, the measurement of high-speed railway channel, and the small-scale fading channel and large-scale fading channel in viaduct and cutting of high-speed railway. Through on-the-spot investigation and measurement, combined with the analysis of propagation mechanism, the classification standard of typical propagation scenarios of high-speed railway is established. According to two typical environments of high-speed railway viaduct and cutting, combined with the measured data and the characteristics of the directional antenna system, the statistical modeling of small-scale fading characteristics is carried out. In addition, the characteristics of path loss and shadow fading in various propagation scenarios are deeply studied. Combined with the channel measurement results, a large-scale propagation model library for high-speed railway communication system design is established to serve the link budget and network design of wireless communication system.

5.1 OVERVIEW

Since the establishment of the first high-speed railway in the world-Shinkansen in Japan in 1964, high-speed railway has been greatly developed in the world, effectively alleviating the traffic demand in densely populated areas. According to the "Mid- and Long-Term Plan of Railway Networks" of China, the operative railway lines in China will be up to 120,000 km long by 2020, including over 16,000 km HSR lines. However, behind the huge social benefits, the huge economic burden brought by high-speed railway to government investment cannot be ignored. According to the financial report of the former Ministry of Railways, the total government investment in high-speed railway projects will exceed 2 trillion yuan by 2020. Excessive economic burden restricts the further development of high-speed railway. A railway control system based on wireless communication is the key

DOI: 10.1201/9781032669793-5

to ensure the safe operation of high-speed railway. However, the cost of wireless communication network construction is not negligible, which accounts for around 40% of the design cost of the whole railway control system. In the actual design of high-speed railway, the cost of a wireless communication network per kilometer exceeds 400,000 yuan. The main consumption is caused by the base station and the corresponding land acquisition, power network layout, logistics support, and other factors. How to design and plan the base station network layout strategy reasonably and effectively and to reduce unnecessary base station investment has gradually become the key to reduce the investment in wireless communication system, improve the efficiency of high-speed railway construction and optimize the construction of high-speed railway system.

With the rapid development of high-speed railway, establishing a reliable and stable railway wireless communication system has gradually become the key to ensure railway safety. The design of wireless communication system cannot be separated from the accurate prediction of wireless signal propagation loss and characteristics in related environments. The high mobility of high-speed railway has greatly changed people's understanding, definition, and classification of wireless propagation scenarios, and put forward higher requirements for the development of wireless propagation models, including the following aspects:

- Pertinence: it can serve the prediction of signal propagation characteristics under the railway characteristic propagation environment;

- Accuracy: it can effectively improve the prediction accuracy of wireless propagation model in related scenarios;

- Integrity: it can describe the change of propagation characteristics of wireless signals in different dimensions of large and small scales, and serve the design requirements of all levels of the system.

Therefore, it is urgent to study the wireless communication channel of high-speed railway in depth, fully understand the wireless communication characteristics of signals in high-speed railway system, and serve the design of high-speed railway wireless communication system more effectively.

In the high-speed railway system, because the actual running speed of trains is often higher than 300 km/hour, a too large rail gradient and curvature will lead to potential safety hazards. Therefore, in order to ensure the smoothness of high-speed railway track, the physical environment along the track has to be artificially transformed in railway construction, for example, viaducts or cutting need to be erected on undulating ground to ensure the smoothness of the track. In addition, infrastructure such as houses along the railway track needs to be stretched away to meet the resettlement of base stations and power equipment along the railway track. The construction of the above-mentioned high-speed railway foundation engineering will produce a large number of characteristic communication environments different from the traditional cellular network environment, among which the most characteristic railway environments are viaduct and cutting.

The emergence of a special propagation environment increases the difficulty of predicting the propagation characteristics of wireless signals in high-speed railway [1,2].

The classification of wireless propagation scenarios is mostly based on the change of a wireless propagation mechanism and the radius size of cells. For example, the Hata model [3] divides three types of environments: urban, suburban, and plain based on the difference of propagation environments; the WINNER II model [4] divides 18 specific scenarios based on the cell radius and application environment; the COST259 model [5] divides 13 specific scenarios based on the difference of multipath structure and cell radius. The classification of high-speed railway wireless propagation scenarios started late. Based on the field investigation of four high-speed railway lines and eight high-speed railway stations in China, ref. [6] put forward the classification standard of high-speed railway radio wave propagation scenarios for the first time, and divided the high-speed railway propagation environment into 18 typical scenarios according to the characteristics of radio wave propagation and user needs. Scenario classification is mainly based on the change of radio wave propagation characteristics and the difference in actual railway wireless service requirements. However, ref. [6] does not provide an accurate propagation model in various scenarios. In addition, there are still two problems in the scenarios division in ref. [6]: First, some communication scenarios (such as desert, sea, etc.) have not yet appeared in the high-speed railway environment; Second, the division of some propagation scenarios is based on the differences of railway control systems and users' needs, rather than the differences of transmission characteristics of wireless communication links. This makes the feasibility of high-speed railway wireless channel modeling using the scenarios classification standard of ref. [6] reduced. Therefore, reasonably extracting the most representative propagation scenarios in high-speed railway environment is the basis of subsequent high-speed railway wireless channel measurement and modeling.

Due to the lack of high-speed railway propagation scenario classification research, wireless channel measurement and modeling research also lack pertinence and systematics. The research and application of wireless channel model of high-speed railway have the following problems:

1. The channel measurement scenarios are unknown

    The measurement of railway propagation channel in refs. [7–9] does not distinguish specific propagation scenarios, but is uniformly classified as suburban scenarios. This makes the influence of many characteristic environments of high-speed railway (such as viaduct and cutting) on radio wave propagation ignored, which reduces the accuracy of subsequent model development.

2. Confusion in model usage

    The Hata model and WINNER II model have been widely used in the design of high-speed railway wireless network due to the lack of targeted channel model guidance [10]. However, the Hata model is not based on the measurement results of high-speed railway channel, and the high-speed mobile scenario in the WINNER II model adopts the propagation model of rural environment, which cannot accurately reflect the propagation characteristics of high-speed railway wireless channel. The confusion of model use directly affects the design quality of wireless network.

3. Fading channel measurements are scarce

Due to the limitation of high-speed railway propagation scenario classification and channel measurement, the characteristics of time-varying fast fading channel in high-speed mobile environment need further study. References [4,11–14] have studied the traditional small-scale fading channels in urban, suburban, and rural scenarios, but the related results cannot be used to estimate the fading channels in railway characteristic scenarios such as viaducts and cutting. Reference [15–18] studies the small-scale fading characteristics of high-speed railway viaduct and cutting environment, and ref. [19] establishes the Ricean distribution *K*-factor model under viaduct scenarios. However, on the one hand, the above research does not consider the influence of the actual high-speed railway directional antenna system on the fading channel [10]; on the other hand, it does not deeply explore the relationship between structural parameters of different scenarios and small-scale fading characteristics. Therefore, the measurement and modeling of a small-scale fading channel in high-speed railway characteristic scenarios need to be studied in depth to better serve the construction and development of high-speed railway wireless communication network.

In the design of high-speed railway communication system, the communication distance between base station and mobile station is affected by the transmission power, antenna gain, channel propagation loss, system cable loss, and the minimum available receiving level of the system. The transmission power, antenna gain, and cable loss depend on the equipment used in the actual network construction, and the minimum available reception level depends on the sensitivity of the receiver used in the system. The above parameters are all determining factors for a given wireless communication system. However, in the process of network design, the estimated value of propagation channel loss depends on the channel model used, so the accuracy of predicting the maximum communicable distance of wireless communication system is closely related to the channel model used. If the model used to predict the actual propagation loss is too large, it will lead to the actual base station spacing in the network design will become smaller, and then increase the network design investment, resulting in serious waste. In fact, this is the main problem existing in the wireless communication design of high-speed railway nowadays, that is, the model used in the system link budget predicts the propagation loss too much [10]. The research on a small-scale fading channel model of high-speed railway is helpful to the development of fading channel simulator in high-speed mobile scenarios and the design of physical layer coding and frame structure of the system; However, the propagation loss model that wireless network coverage prediction and link budget depend on depends more on the large-scale propagation characteristics of the channel, namely path loss, and shadowing.

The research on the measurement and modeling of large-scale propagation characteristics of wireless channels has a long history. References [3,5,20–28] introduce the research on measurement and modeling of large-scale propagation characteristics of channels in different frequency bands and different scenarios under traditional cellular networks. However, due to the special propagation environment and frequency band in high-speed railway, the above results cannot be used in the design of high-speed railway wireless communication system. In recent years, the research on large-scale characteristics of high-speed railway channel has gradually attracted people's attention. Table 5.1 summarizes the representative research work on large-scale characteristics of high-speed railway channels in recent years. It is found that the related research in this field has the following shortcomings.

1. Most studies focus on viaduct scenarios [19,29–33]

   Although viaduct is the most typical characteristic scenario in high-speed railway environment, according to the statistical proportional distribution of each scenario in high-speed railway, the other six scenarios still occupy a high proportion, and the lack of the corresponding propagation model will bring great obstacles to system link budget and network design. Only by establishing a perfect and accurate propagation model base can we really meet the needs of high-speed railway wireless network design.

2. Most of the measurements are only performed without the use of real high-speed railway communication systems [4,8,19,31–35] (930 MHz carrier frequency, directional base station antenna, on-board receiving antenna and high-speed train)

TABLE 5.1   Summary of Investigation on Shadow Fading in HSR Environments

References	Frequency	Scenarios	Use GSM System	Path Loss	Shadow Fading Distribution Characteristics	Shadow Fading Auto- Correlation	Shadow Fading Cross- Correlation
[29]	930 MHz	Viaduct	Yes	Has been studied	Unstudied	Unstudied	Unstudied
[30]	930 MHz	Viaduct	Yes	Unstudied	Has been studied	Has been studied	Unstudied
[31]	2.35 GHz	Viaduct	No	Unstudied	Has been studied	Has been studied	Unstudied
[19]	2.35 GHz	Viaduct	No	Has been studied	Has been studied	Unstudied	Unstudied
[32]	2.35 GHz	Viaduct	No	Has been studied	Has been studied	Unstudied	Unstudied
[33]	2.6 GHz	Viaduct	No	Has been studied	Has been studied	Unstudied	Unstudied
[34]	2.35 GHz	Cutting	No	Has been studied	Has been studied	Unstudied	Unstudied
[4]	5.2 GHz	Suburb	No	Has been studied	Has been studied	Unstudied	Unstudied
[35]	400 MHz	Rural	No	Unstudied	Unstudied	Unstudied	Unstudied
[8]	320 MHz	Rural	No	Has been studied	Unstudied	Unstudied	Unstudied
[36]	930 MHz	Station	Yes	Has been studied	Has been studied	Unstudied	Unstudied

The characteristic antenna systems of high-speed railway and many character-istic infrastructures around the railway, such as power lines, towers, rails, etc., are important components of the characteristic transmission environment of high-speed railway: frequency directly affects the dissipation loss of wireless signal transmission [37]; the antenna characteristics of the base station greatly affect the change of path loss [3]; Vehicle antenna is easily affected by large metal car body [38]; However, power lines, towers, and other facilities in the Fresnel zone along the railway directly affect the absorption and diffraction loss of wireless signals [39]. In channel measurement, ignoring the characteristic factors in the actual system will not obtain the propagation channel data reflecting the real characteristics of the communication system, which will reduce the accuracy of the model. Therefore, it is necessary to carry out channel measurement based on a real high-speed railway communication system and establish a real channel model for high-speed railway system design.

3. Most of the works have not studied the characteristics of shadow fading.

At present, the distribution characteristics and autocorrelation characteristics of shadow fading in various scenarios of high-speed railway need to be studied, while the cross-correlation characteristics of shadow fading of adjacent base stations are still insufficient. The former directly affects the analysis of handover delay and "ping-pong effect" of high-speed railway system [40,41], while the latter is helpful to design more efficient handover rules [42,43]. Therefore, the shadow fading charac-teristics in the characteristic environment of high-speed railway need further study.

5.2 SCENARIO CLASSIFICATION AND CHANNEL MEASUREMENT

5.2.1 Scenario Classification Standard of High-Speed Railway

The propagation characteristics of radio waves depend on the geometric structure and electromagnetic characteristics of the communication environment. Before modeling the wireless channel, it is necessary to classify the working environment of the communica-tion system in detail and accurately. However, this work has not been paid attention to the research of high-speed railway wireless communication in recent years. The main reason lies in the lack of targeted scenarios investigation and rigorous, scientific, and accurate sce-narios evaluation criteria at this stage. In the existing models, the scenario classification of radio wave propagation mostly depends on the physical analysis of radio wave propagation characteristics in the scenarios and the subjective perception of models, such as the Hata model [3], WINNER II model [4] and COST 259 model [5]. In order to effectively serve the establishment of the wireless propagation model of high-speed railway and meet the needs of network planning and optimization of high-speed railway at present, based on refs. [6,44], the high-speed railway propagation scenarios are screened and defined again, and a set of targeted scenario division standards suitable for existing high-speed railway network construction is put forward.

The classification of high-speed railway propagation scenarios follows the following principles:

- The division of propagation scenarios starts from the basic mechanism of radio wave propagation [45], which is mainly based on the differences of wireless signal propagation characteristics and multipath structure distribution in various scenarios.

- Propagation scenarios are exclusive: any two types of propagation scenarios do not overlap each other in physical space. This can effectively avoid the confusion of channel model usage in different scenarios.

- Scenarios division does not consider compound scenarios for the time being. Because the propagation environment along the high-speed railway is complex and changeable, there are often many combinations of different propagation environments in a single community. In the process of channel measurement and modeling, this book only uses cells whose typical characteristics of propagation scenarios in a single cell exceed 80% of the cell region for channel model development. In this way, the purest radio wave propagation characteristics in various scenarios can be effectively extracted for analysis and modeling.

- Scenarios classification is based on topography and scatterer distribution in the radial effective region of railway track. The radial effective region of railway track is the equivalent average distance of the main lobe of the antenna in the longitudinal coverage region on both sides of the railway track, which is calculated according to the horizontal 3 dB beam width of the base station antenna of high-speed railway. For a typical high-speed railway system, the radial effective region of the rail is around 500 m long rectangular region on both sides of the rail. By defining the radial effective region of railway track, the influence of main reflectors and scatterers in the environment on wireless propagation can be highlighted, and the scientific and accurate classification of propagation scenarios under a directional antenna system can be improved.

It is worth noting that, similar to the principle of scenarios classification in the COST259 model [5], the scenario classification standard proposed in this book is dedicated to covering generally "typical" scenarios of high-speed railway environment, and has the feasibility of channel modeling under each scenario, instead of including generally possible scenarios. For example, deserts, sea, marshalling yards, and other scenarios are no longer considered in the classification standard of this book, because such scenarios have not yet appeared in the actual lines of high-speed railway or rarely appear. Through the above model screening, this book effectively and reasonably reduces the number of high-speed railway radio wave propagation scenarios, and at the same time can meet the design requirements of high-speed railway wireless communication system.

Finally, the division standard of high-speed railway wireless communication scenarios proposed in this section includes seven types of scenarios. Based on the description of communication scenarios in reference to standardized manuscripts [3–5], the specific definitions of seven types of high-speed railway communication scenarios are as follows:

1. Urban

   This scenario refers to typical large and medium-sized urban, regions with dense population and developed industry and commerce. In the urban scenarios, the railway track surface can be parallel to the ground surface or located on the viaduct of 5~20 m. There are 5~20 floors (10~40 m higher than the derailment surface) in 80% of the residential regions in the radial effective regions on both sides of the rail surface. High-rise buildings in urban scenarios will lead to a large number of strong reflection and scattering paths. This kind of reflection component has a small diffraction loss to any obstacle because of its high secondary emission point, which has a great influence on the signal strength at the receiving end.

2. Suburbs

   This scenario refers to typical small cities, townships, towns, rural, and the countryside in uneven and non-open regions. In suburban scenarios, the railway track surface is mostly parallel to the ground surface. There are 1~5 floors of buildings and vegetation of similar height in the radial effective regions on both sides of the rail surface and 60%~80% of the residential regions in the longitudinal direction. There are proper and evenly distributed reflectors and scatterers in suburban scenarios. Direct radius, ground reflection radius, and environmental reflection radius all occupy a certain proportion.

3. Rural

   This scenario is similar to the plain open land in the Hata model, especially the open country area on the typical flat surface. In rural scenarios, the track surface is mostly parallel to the surface. In the radial effective regions on both sides of the rail surface, 80% of the communities in the longitudinal direction have no buildings, but only a small number of low crops and vegetation (less than 2 m). There are few reflectors and scatterers in rural scenarios, and direct waves and ground reflected waves are dominant.

4. Viaduct

   This scenario refers to the region where the track surface is placed above the viaduct with a height of 10~30 m in non-urban, non-mountainous regions, and non-river regions. In viaduct scenarios, in the radial effective regions on both sides of the rail surface, 80% of the residential regions in the longitudinal direction are low-rise houses with some trees, and the towers are 0~10 m higher than the bridge deck. Viaduct deck will block the backscattered waves of most surface scatterers. However, a flat bridge deck will produce strong reflected waves on the bridge deck. Trees and houses above the bridge deck will cause near-end reflection at the receiving end, resulting in a multipath clustering phenomenon.

5. Cutting

This scenario refers to the U-shaped groove region dug in an uneven zone to ensure the smoothness of rail surface. The steep walls on both sides of the cutting are mostly symmetrical, with an inclined slope of 30° and a depth of 2~20 m, and the surface is mostly covered by vegetation. Outside the cutting, there are mostly suburban and rural environments with sporadic small soil slopes. The deep trench structure of the cutting makes it difficult for the backscattered waves outside the cutting to reach the receiving end, while the steep walls on both sides of the cutting will cause a large number of near-end backscattered waves at the receiving end.

6. Station

This scenario refers specifically to the large, medium, and small passenger stations appearing in the railway line. Above the station, there are awnings with a length of 400~800 m, a width of 100~500 m, and a height of 50~80 m. Large and medium-sized station awnings are closed, while small station awnings are mostly semi-closed awnings at the platform. Most of the base stations stand at 100~500 m outside the canopy. The station roof will cause extra diffraction loss to the transmission of radio waves. Its internal closed structure will lead to dense reflection effect in some regions.

7. Rivers

In this scenario, there are large lakes (more than 1 km²) or rivers with a width of 50~200 m passing through the rails from below in the radial effective regions on both sides of the rails (at this time, the rails are mostly placed on viaducts). Most of the two sides of the railway track are typical suburban and rural environments. The water surface will cause a large number of specular reflections. At the same time, because the absorption loss of water surface is different from that of the surface, its reflection coefficient is larger, and the wireless loss is greatly affected by it.

Figure 5.1 shows the typical scenarios of the above seven types of high-speed railway, from which it is found that that the four types of scenarios, viaduct, cutting, station and river, are significantly different from the traditional cellular scenarios. Figure 5.2 shows the proportion of each scenario of China's "Zhengzhou-Xi'an" high-speed railway, among which the characteristic scenarios of high-speed railway, such as viaduct, cutting, and station, account for a much higher proportion than the scenarios of urban and rural in cellular cells. This also reflects the necessity of defining typical communication scenarios of high-speed railway. It should be pointed out that the above classification standard of communication scenarios does not cover generally scenarios in high-speed railway environment, but it is sufficient to meet the design requirements of high-speed railway network today. The propagation model established according to this standard can effectively improve the accuracy of wireless channel simulation of high-speed railway at present. Please note that the above description of propagation environment is only used for qualitative analysis of propagation channel. In the channel measurement and modeling later, the specific

(a) From left to right: Urban, Suburban, Rural

(b) Viaduct

(c) Cutting

(d) Station

(e) River

FIGURE 5.1   High-speed railway communication scenario.

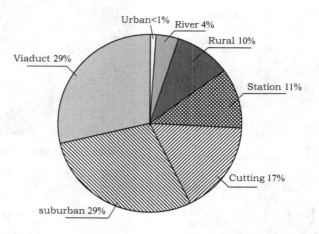

FIGURE 5.2   Proportion of each scenario of China's "Zhengzhou-Xi'an" high-speed railway

test and environmental structure parameters in individual characteristic scenarios (viaduct and cutting) will be given in detail. The general situation of various scenarios is introduced in this chapter, and the related contents will not be repeated in other chapters.

In addition, the above propagation model classification standard does not include tunnel scenarios. This is because in today's high-speed railway communication system, the wireless coverage of tunnel scenarios mainly adopts leaky cable mode [46], rather than cell coverage mode based on base station antenna. Because the leaky cable coverage mode has high and stable signal coverage quality, the wireless system design in this kind of scenario does not depend on the traditional channel model. Therefore, in the above scenario classification and subsequent channel modeling, the tunnel scenario is not considered in this section. References [47–49] are available for wireless coverage in tunnels in leaky cable mode, and related statistical measurements of channels in tunnels are available in refs. [50–52].

5.2.2 High Speed Railway Test System and Channel Measurement

In order to study the wireless channel of high-speed railway, field measurements were carried out on "Zhengzhou-Xi'an" and "Beijing-Shanghai" high-speed railway passenger dedicated lines in China. The test system and high-speed railway field measurement will be described in detail below.

5.2.2.1 High-Speed Railway Channel Test System

In order to fully reflect the channel characteristics of high-speed railway wireless communication, the standard GSM-R system used in practice is used to measure the propagation channel. Each module of the high-speed railway test system is shown in Figure 5.3.

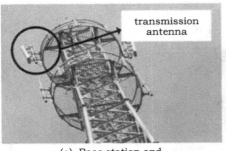

(a) Base station and
APX86-906515S-CT0 transmission antenna

(b) High speed trains used for measurement

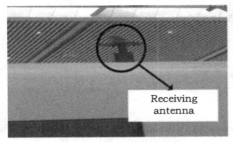

(c) 900/2400-HBNT receiving antenna

(d) Willtek 8300 Griffin field strength tester

FIGURE 5.3   High speed railway test system.

The specific description is as follows:

1. Launching pad:

    The GSM-R actual system base station is used as transmitting station. The base station tower is usually located 15 m away from the longitudinal direction of the rail surface, and the base station antenna is 20~40 m away from the rail surface. In the measurement, the broadcast control channel signal of the 930 MHz frequency band is used as the transmission signal, and the output power of the signal after passing through the amplifier at the transmitter is 43 dBm. The transmitter feeder is connected to the 17 dBi APX86-906515S-CT0 directional antenna. The 3 dB beam widths of the antenna in horizontal and vertical directions are 65° and 6.8°.

2. Receiving station:

    A 900/2400-HBNT horizontal omni-directional antenna with a height of 30 cm is used as the receiving antenna. The antenna gain is 4 dBi and the vertical 3 dB beam width is 80°. The antenna is placed on the top of the front compartment of the high-speed train. The high-speed train conductor used in the test is 204 m long, 3.8 m high, and 3.3 m wide. The receiving antenna is connected to the Willtek 8300 Griffin field strength tester to capture and store the signal power in real time. Distance sensors are installed on the wheels of high-speed trains, which are connected with Willtek 8300 Griffin field strength tester to record GPS information of measured data in real time. With the support of GPS system, Willtek 8300 Griffin field strength tester can support two modes: distance trigger and time trigger. The measurement adopts distance trigger mode, in which the tester can sample data at a fixed distance interval, and this process is not affected by the change of actual running speed of trains.

In addition, Willtek 8300 Griffin field strength tester supports the frequency sweep measurement mode, which can receive transmitted signals from different frequency points of adjacent base stations at the same time. In the high-speed railway wireless communication system, the frequency difference between the signals transmitted by adjacent base stations is 12 MHz. Because the bandwidth of GSM-R signal is only 0.2 MHz, the interference between adjacent base stations can be ignored. Therefore, in the channel measurement, the signals of adjacent base stations along the high-speed railway can be recorded synchronously in real time, which serves the research of channel shadow fading correlation.

5.2.2.2 Directional Base Station Antenna of High-Speed Railway

High-gain directional base station antenna is adopted in the actual construction of high-speed railway wireless communication system. Figure 5.4 shows the schematic diagram of high-speed railway directional base station antenna system, where z is the included angle between antenna and base station tower; $\alpha_b = 6.8°$ is the vertical 3 dB beam width of the base station antenna, and the main lobe of the base station antenna points along the railway track to realize the strip coverage of the railway line; h_t and h_r denote the heights of the base station and mobile station antennas from the rail surface respectively. In the actual high-speed railway survey, h_t ranges from 20 to 40 m and α_1 ranges from 2° to 7° according to

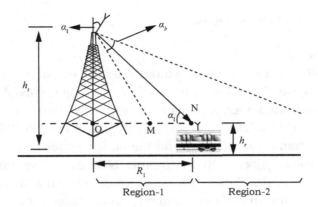

FIGURE 5.4   Schematic diagram of antenna system of high-speed railway base station.

the requirements of different district configuration and coverage. The base station antenna is placed at a longitudinal distance of 15 m from the rails. In addition, the solid lines in Figure 5.4 indicate the main radiation direction of the antenna, that is, the direction of the maximum gain in the main lobe pattern. The gain distribution in the main lobe region of antenna pattern is relatively uniform. However, in the sidelobe region, the gain far away from the main radiation direction of the antenna is small and fluctuates greatly.

A directional base station antenna has a great influence on the coverage of railway wireless community. Figure 5.5 shows an example of wireless signal reception power measurement results in three cells in a high-speed railway suburban scenario, in which the small-scale fading characteristics of the received power are pre-eliminated. In the figure, the distance of 0 m indicates the position of the base station tower. From Figure 5.5, it can be found that the received signal strength near the base station (i.e., the sidelobe coverage region of the base station antenna) is small and fluctuates greatly, and the radio wave propagation path loss in this region no longer follows the distance index model in the traditional cellular network [37].

This is due to the irregular fluctuation of sidelobe gain of base station antenna around 20 dB. Therefore, it is necessary to distinguish the main lobe and sidelobe coverage regions of base station antenna when modeling the channel of high-speed railway directional antenna system. For the convenience of subsequent channel modeling the coverage region of the antenna is defined here as Region 1 and Region 2 which are explained as follows:

- Region 1: At this time, the train is located at the bottom of the base station tower, and the received signal is mainly affected by the sidelobe gain of the directional antenna of the base station. In Region 1, although there is a direct radius, the received signal still suffers a certain degree of loss at different positions due to the low sidelobe gain of directional antenna and irregular fluctuation of the sidelobe pattern. As shown in Figure 5.4, the size of Region 1 is R_1.

- Region 2: As shown in Figure 5.4, the train leaves the bottom of the base station tower in this zone, and the received signal is mainly affected by the main lobe gain of the directional antenna of the base station.

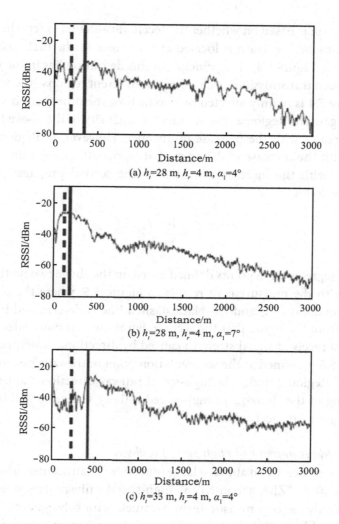

(a) h_t=28 m, h_r=4 m, α_1=4°

(b) h_t=28 m, h_r=4 m, α_1=7°

(c) h_t=33 m, h_r=4 m, α_1=4°

FIGURE 5.5 Propagation region discontinuity verification based on test data from 3 base stations.

In order to reasonably estimate the boundary points between Region 1 and Region 2 (the propagation segmentation points), the geometric derivation of base station antenna structure combined with empirical measurement results is adopted. As can be seen from Figure 5.4, there are two possible ways to define propagation segmentation points.

The first definition is based on the geometric relationship between the critical positions of the mainlobe and sidelobe of the antenna, that is, the horizontal propagation distance corresponding to the edge of the sidelobe of the antenna (that is, the distance between O and M in Figure 5.4).

$$R_1 = \frac{h_t - h_r}{\tan\left(\alpha_1 + \dfrac{\alpha_b}{2}\right)} \tag{5.1}$$

The second definition is based on whether the receiving antenna enters the main radiation region of the mainlobe, the train is located at the center of the main lobe (the distance between O and N in Figure 5.4). The principle of this definition lies in the judgment of the relationship between transmission mechanisms in different regions. The wireless propagation loss in Figure 5.4 is mainly affected by two factors: the propagation loss in free space and the antenna gain. In Region 2, the attenuation intensity of the above two propagation mechanisms increases with the increase of distance. However, in Region 1, the antenna gain increases with the increase of distance, that is, the attenuation intensity of wireless signals decreases with the increase of distance. The second propagation segmentation point is defined as follows:

$$R_1 = \frac{h_t - h_r}{\tan \alpha_1} \tag{5.2}$$

The propagation segmentation points defined based on the above two methods are marked and compared with the measurement results in Figure 5.5, where the black solid line is determined by equation (5.2) and the black dashed line is determined by equation (5.1). It can be found from the figure that the second definition can more effectively reflect the sudden change of received signal strength caused by directional antenna. Thus, the position N in Figure 5.5 is defined as the segmentation point of the wireless propagation channel Region 1 and Region 2 under the high-speed railway directional antenna system. The channel modeling of the subsequent high-speed railway scenarios will be carried out in Region 1 and Region 2 respectively.

5.2.2.3 Channel Measurement of High-Speed Railway

Using the test system described above, a large number of channel measurements have been carried out in China's "Zhengzhou-Xi'an" high-speed railway passenger dedicated line (please note that only the core measurement parameters for subsequent analysis and modeling are introduced here, and the basic theory of channel measurement methods can be referred to refs. [53,55]). Measurements are based on different sampling interval patterns and can be divided into the following two categories:

- 53 cm sampling interval channel measurement: With the support of GPS system, Willtek 8300 Griffin field strength tester is set to distance trigger mode, and the distance sampling interval is 53 cm. A large number of channel measurements with 53 cm sampling intervals have been carried out in all seven scenarios of high-speed railway. Table 5.2 summarizes the number of cells used for channel measurement in various scenarios. Except for urban scenarios, hundreds of measurements have been carried out in different communities in the other six types of scenarios to maximize the accuracy and universality of the model. The test data with 53 cm sampling interval can be used to study and model the large-scale characteristics of the channel.

TABLE 5.2   Number of Measurement Cells in Various
Environments in the High-Speed Railway Channel Measurement

Environment	Number of Cells
Urban	66
Suburb	1,500
Rural	808
Viaduct	2,144
Cutting	638
Station	772
River	248

- 10 cm sampling interval channel measurement: Willtek 8300 Griffin field strength tester is set to distance trigger mode with the support of GPS system, and the distance sampling interval is 10 cm. Channel measurements with 10 cm sampling intervals are carried out in high-speed railway viaduct and cutting scenarios, which are used to study and model the small-scale characteristics of the channel. Limited by the specific test conditions, the channel measurement with 10 cm sampling interval was not carried out in the other five scenarios except viaduct and cutting scenarios. See the following contents of this chapter for the specific number of test cells and measurement parameters in viaduct and cutting scenarios. The test data with a 10 cm sampling interval can be used to study and model the small-scale characteristics of the channel.

In Section 5.3 of this chapter, a large-scale propagation model library of high-speed railway is established by using the test data of 53 cm sampling interval in seven railway scenarios to meet the needs of system link budget and network planning. Section 5.4 will use the 10 cm sampling interval channel test data in viaduct and cutting scenarios to carry out in-depth research and modeling of small-scale fading characteristics, so as to meet the needs of channel time-varying characteristic estimation and fast fading channel simulation in high-speed railway high mobility scenarios.

5.3 LARGE-SCALE CHANNEL MODELING

5.3.1 Path Loss

This section introduces the large-scale propagation model library of high-speed railway established by using 53 cm sampling interval test data in seven scenarios of high-speed railway to serve the requirements of link budget and network planning of wireless communication system [56]. In order to meet the requirements of high-speed railway characteristic communication system, the path loss model is established in Region 1 and Region 2 shown in Figure 5.4.

5.3.1.1 Antenna Gain Calibration and Data Preprocessing

In order to establish an accurate path loss model, it is necessary to calibrate and compensate the antenna pattern to remove the influence of pattern fluctuation on the estimated path loss [57]. This channel model can be used to predict the channel loss under different antenna systems and extend the application range of the model.

Firstly, the original measurement data at 53 cm sampling interval are averaged by using 40X wavelength sliding window, and the obtained local statistical mean value is the large-scale sample value. Local logarithmic domain large-scale loss samples can be expressed as:

$$\hat{PL}(d) = P_t - P_r(d) + G_v(d, \alpha_1) - L_{loss} \tag{5.3}$$

where $P_t = 43$ dBm is the transmitting power of the high-speed railway base station; P_r is the average value of the local received power obtained by statistical averaging in the 40-fold wavelength window; $L_{loss} = 9.05$ dB is the loss of the feeder and device connector in the high-speed railway test system; $G_v(d, \alpha_1)$ is the total of the transmitting antenna and the receiving antenna Vertical gain:

$$G_v(d, \alpha_1) \left(G_{v,\text{RX}} \tan^{-1}\left(\frac{\Delta h}{d}\right) \right) + G_{v,\text{TX}}\left(\alpha_1 - \tan^{-1}\left(\frac{\Delta h}{d}\right) \right), \alpha_1 - \tan^{-1}\left(\frac{\Delta h}{d}\right) \geq 0$$

$$\tag{5.4}$$

$$G_v(d, \alpha_1) \left(G_{v,\text{RX}} \tan^{-1}\left(\frac{\Delta h}{d}\right) \right) + G_{v,\text{TX}}\left(360 + \alpha_1 - \tan^{-1}\left(\frac{\Delta h}{d}\right) \right), \alpha_1 - \tan^{-1}\left(\frac{\Delta h}{d}\right) < 0$$

where $G_{v,\text{TX}}(\cdot)$ and $G_{v,\text{RX}}(\cdot)$ respectively denote the vertical gain of the transmitting antenna and the receiving antenna; α_1 is the angle between the antenna and the base station tower in Figure 5.4; $\Delta h = h_t - h_r$ is the height difference between the transmitting antenna and the receiving antenna. Because the horizontal beam of the base station antenna is wide, the horizontal gain of the antenna remains basically unchanged in the linear coverage of the railway residential region, so the change of the horizontal dimension pattern gain of the antenna is not considered in the subsequent analysis.

The $\hat{PL}(d)$ samples estimated by equation (5.3) contains both large-scale path loss and shadowing information. LS regression fitting functions with $\hat{PL}(d)$ varying with distance (path loss PL); samples of $\hat{PL}(d)$ randomly fluctuate around the PL regression model (shadow fading X_{LS}) [37]. In the following sections, the local large-scale path loss and shadowing samples will be sequentially stripped from equation (5.3) for statistical modeling.

5.3.1.2 Path Loss Modeling in Region 1

In the Region 1 shown in Figure 5.4, because the receiving station is close to the base station, the intensity of the far-end backscattered wave is far lower than the direct radius and the ground reflection radius, and the received signal is mainly affected by the energy of the direct radius and the ground reflection radius. The above phenomenon is true in seven types of high-speed railway scenarios. Therefore, the conventional two-ray model [37] is used to predict the change of path loss within Region 1 in an arbitrary high-speed railway scenario. The path loss of the two-ray model can be expressed as:

$$PL_{\text{Two-Ray}} = \left(\frac{\lambda}{4\pi}\right)^2 \left| \frac{e^{-jk_\lambda d_d}}{d_d} + \zeta \frac{e^{-jk_\lambda d_r}}{d_r} \right|^2 \tag{5.5}$$

d_d is the propagation distance of the LOS between the transmitting station and the receiving station; d_r is the propagation distance of the ground reflection path; k_λ is the wave number; ζ is the ground reflection coefficient, which is taken as −1 in common environments [38]. The two-ray model can be effectively used in propagation environments where the LOS and the ground-reflected path dominate, while having low computational complexity.

Figure 5.6 shows examples of received signal power measurements estimated based on equation (5.3) under the same type of cell (in Figure 5.4, $h_t = 33\,\text{m}, \alpha_1 = 5°$) within a high-speed railway suburban scenarios Region 1, and the comparison results of the two-ray propagation model and the free-space propagation model [58]. In the two-ray model, due to the superposition of the amplitude and phase of the direct radius and the ground reflection radius at different positions, the multipath energy appears coherence and cancellation at different positions, so that the total power of the received signal exhibits oscillation phenomenon shown in Figure 5.6. As can be seen from Figure 5.6, the received power variation in Region 1 after antenna gain calibration approximately obeys a two-ray model. In addition, Figure 5.6 compares the prediction effects of antenna gain calibration with and without antenna gain calibration using equation (5.4). The results show that only after antenna gain calibration using equation (5.4) can accurate prediction results be obtained. Note that the prediction accuracy of the two-ray model has also been verified in the remaining six scenarios, and only Figure 5.6 is used here as an example of the verification results.

In order to further verify the accuracy of the two-ray model in Region 1, the difference between the average value of path loss in each cell and the predicted average value of the two-ray model is defined as an error factor Δ:

$$\Delta(i) = \overline{\text{PL}(i)} - \overline{\text{PL}_{\text{Two−Ray}}(i)} \tag{5.6}$$

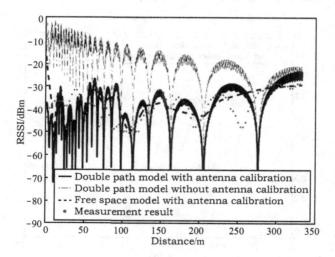

FIGURE 5.6 Sample measurement of Region 1 in suburban scenarios of high-speed railway ($h_t = 33\,\text{m}$, $\alpha_1 = 5°$).

where i is the high-speed railway community number. Because of the small range of Region 1, the wireless signal strength will not change obviously with the increase of distance, but is more affected by the coherence and cancellation of direct radius and ground reflection radius in the short range of dual-path model and the fluctuation of the sidelobe pattern of base station antenna. Therefore, the difference between the statistical mean value and the measurement result of the prediction model in Region 1, that is, the error factor Δ, can effectively reflect the accuracy of the prediction model. Figure 5.7 shows the CDF statistics for the error factor delta in the high-speed railway Region 1. It is found that from the figure that the measured mean value of the path loss in each cell after calibrating the antenna pattern is very close to the predicted mean value of the two-ray model, and the mean error is only ± 0.43 dB. The results show that, based on the calibration of antenna sidelobe pattern, the variation of propagation characteristics of large-scale channel in Region 1 in various scenarios of high-speed railway obeys the two-ray propagation model.

5.3.1.3 Path Loss Modeling in Region 2

In this section, the original large-scale channel loss samples obtained by equation (5.3) are used to analyze the path loss PL in Region 2 under the seven types of high-speed railway scenarios($\hat{\text{PL}}$ extracted from equation (5.3) changes with the distance d LS regression function) for statistical modeling.

The path loss PL increases linearly with the increase of logarithmic distance $\log d$:

$$PL = PL_0 + 10n \log d \tag{5.7}$$

A large number of refs. [37,39,40] show that the constant term PL_0 and path loss exponent n in equation (5.7) are affected by carrier frequency, antenna height, and environmental parameters. How to simulate the change of PL_0 n reasonably becomes the key to path loss modeling. In statistical channel modeling, adding adjustable parameters of the model can

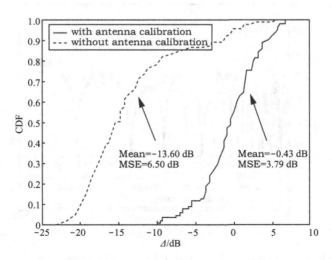

FIGURE 5.7 Error factor CDF in high-speed railway Region 1.

improve the goodness of fit of multiple regression model to actual measured data. However, with the increase of the number of parameters, the degree of freedom of the model will also increase, which makes the convergence of the parameters in the model decrease [59], and then affects the scientific use of the model in practical engineering. In addition, too much introduction of environmental structural parameters in high-speed railway will also bring extra statistical modeling errors and reduce the accuracy of the model. Therefore, it is very important to establish a large-scale path loss model of high-speed railway by reasonably selecting the description of model parameters and the change of PL_0 and n. Since the carrier frequency (930 MHz) and the antenna height (4 m) of the GSM-R system are fixed, the relative height h_b of the base station antenna from the rail surface is introduced as an adjustable parameter for statistical modeling. The change of h_b can effectively reflect the change of the relative position between Fresnel zone [60] and the surrounding Reflectors and Scatterers of the communication link in the cell, and then it can be reasonably used to estimate and model the large-scale propagation loss.

Because the empirical formula based on Hata model [3] is widely used in high-speed railway network planning and related software tools, in the statistical modeling of this book, the statistical path loss model under high-speed railway scenarios is also established based on the basic structure of Hata model. The Hata model includes three kinds of communication scenarios: urban, suburban, and rural. Among them, the path loss model of suburban and rural scenarios is obtained by adding correction factors on the basis of Hata urban scenarios model. In the Hata model, the path loss model of large urban can be expressed as:

$$PL_{Hata}(dB) = 74.52 + 26.16\log f - 13.82\log h_b - 3.2(\log(11.75h_m))^2 + [44.9 - 6.55\log h_b]\log d$$

$$(5.8)$$

where f is the signal carrier frequency and the unit is MHz; In the Hata model, h_b and h_m denote the effective heights of the base station and mobile station antennas.

According to the above analysis, based on equation (5.8), the path loss prediction model under high-speed railway scenarios is proposed in the form of adding correction factors:

$$PL_{HSR}(dB) =$$

$$\Delta_1 + 74.52 + 26.16\log(f = 930) - 13.82\log h_b - 3.2(\log(11.75h_m))^2 + [44.9 - 6.55\log h_b + \Delta_2]\log d$$

$$(5.9)$$

where, h_b and h_m denote the heights of the base station and mobile station antennas "from the rail surface" respectively, and the unit is m. Δ_1 and Δ_2 are model correction factors, Δ_1 is used to correct the path loss model constant item to improve the model goodness of fit; Δ_2 is used to correct the path loss index item in the model, because a lot of previous work has shown that [6,61–63], the path The loss index is affected by the high-speed railway characteristic scenarios. In addition, in order to facilitate subsequent modeling of high-speed railway wireless channels in different frequency bands, the original frequency factor of the Hata model is retained in equation (5.9).

In this section, the linear regression method based on a large number of community test samples is used to obtain the LS regression fitting value of the path loss (PL) in each community in each type of high-speed railway scenario and the constant term and path loss index term in equation (5.9). Through the large-scale channel measurement under the sampling interval of 53 cm in Section 5.2, the statistical values of Δ_1 and Δ_2 in various high-speed railway scenarios are obtained. Then through joint analysis with the height information of the base station in each cell, it is found that Δ_1 and Δ_2 are affected by the height h_b of the base station antenna in the logarithmic domain. To this end, equation (5.10) is used to carry out statistical modeling of Δ_1 and Δ_2 in different high-speed railway scenarios.

$$\begin{cases} \Delta_1 = p_{\Delta_1} \cdot \log h_b + q_{\Delta_1} \\ \Delta_2 = p_{\Delta_2} \cdot \log h_b + q_{\Delta_2} \end{cases} \tag{5.10}$$

where p_{Δ_1}, q_{Δ_1}, p_{Δ_2}, and q_{Δ_2} are undetermined coefficients, which can be obtained by LS regression fitting of a large number of cell measurement samples. If there is no obvious increasing or decreasing relationship between Δ_1, Δ_2, and h_b, p_{Δ_1} and p_{Δ_2} are set to 0, and only q_{Δ_1} and q_{Δ_2} (the statistical mean of a large number of cell measurements in the same scenario) are used as statistical models.

Figures 5.8 and 5.9 respectively show the estimated Δ_1 and Δ_2 based on a large number of cell measurements in the high-speed railway station scenario where it is found that that Δ_1 increases with an increase in $\log h_b$; Δ_2 does not change with $\log h_b$. The coefficients in equation (5.10) are obtained according to LS regression fitting, respectively, as the final model of correction term (where $p_{\Delta_2}=0$). In other high-speed railway scenarios, a similar method is used to estimate Δ_1 and Δ_2 in each scenario. By combining the statistical models

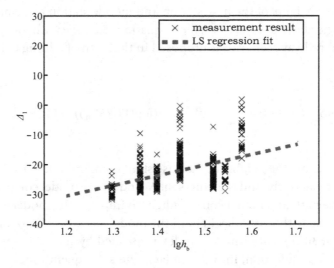

FIGURE 5.8 Example of log domain Δ_1 estimation and LS regression fitting curve in high-speed railway station scenario.

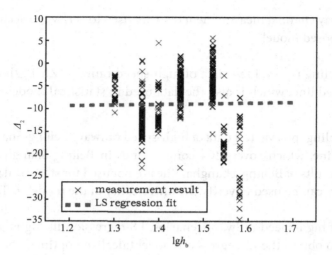

FIGURE 5.9 Example of log domain Δ_2 estimation and LS regression fitting curve in high-speed railway station scenarios.

TABLE 5.3 Modification Factor of the Environmental Path Loss Model of High-Speed Railway

Environment	Correction Factor
Urban	$\Delta_1 = -20.47$ $\Delta_2 = -1.82$
Suburb	$\Delta_1 = 5.74\log h_b - 30.42$ $\Delta_2 = -6.72$
Rural	$\Delta_1 = 6.43\log h_b - 30.44$ $\Delta_2 = -6.71$
Viaduct	$\Delta_1 = -21.42$ $\Delta_2 = -9.62$
Cutting	$\Delta_1 = -18.78$ $\Delta_2 = 51.34\log h_b - 78.99$
Station	$\Delta_1 = 34.29\log h_b - 70.75$ $\Delta_2 = -8.86$
River	$\Delta_1 = 8.79\log h_b - 33.99$ $\Delta_2 = -2.93$

of Δ_1 and Δ_2 with equation (5.9), the statistical modeling of path loss in various scenarios of high-speed railway is realized. Table 5.3 summarizes the modeling results of the correction factor of the path loss model in equation (5.9) under different environments of high-speed railway. It is found that from this the expression form of the correction factor is similar to the Hata model, which is helpful for the popularization and application of this model in the design of high-speed railway wireless network.

5.3.1.4 Model Validation

In order to verify the accuracy of the path loss model in Table 5.3, two indicators, the coefficient of determination R^2 and RMSE, are used. The following two sets of different high-speed railway channel measurement data are used to verify the accuracy and universality of the proposed model.

- 53 cm sampling interval test data of high-speed railway "Zhengzhou-Xi'an" passenger dedicated line, which is also the data used in statistical modeling of path loss in Table 5.3.

- 53 cm sampling interval test data of high-speed railway "Beijing-Shanghai" passenger dedicated line, which covers 104 communities in Beijing-Shanghai line. The measurement results of Beijing-Shanghai line are not used for statistical modeling of path loss, so they can be used to verify the universality of the model in Table 5.3.

For each type of high-speed railway scenarios, LS regression fitting is performed on the measured data to obtain the LS regression model (idealized optimal channel model), and the R^2 and RMSE of the LS regression model are estimated. Then use the test data of the cell to estimate the R^2 and RMSE of the path loss model, Hata model, ITU-R vehicle mobility scenario model [37], 3GPP model [37] and WINNER high-speed mobility scenario model in Table 5.3. Due to the large number of test plots, the CDF statistical characteristics of the above indicators in all plots were compared to test the overall performance of the model.

Figures 5.10 and 5.11 show the statistical distribution comparison of the coefficient of determination R^2 of different models in the tests of the "Zhengzhou-Xi'an" and "Beijing-Shanghai" high-speed passenger dedicated railway lines. The "LS optimal regression fitting" curve in the figure indicates the prediction performance of the LS regression fitting curve of the measurement results in each plot (the optimal prediction performance under ideal conditions). It is found from the figure that the performance of the proposed model is better than the other four types of existing standard models, and is very close to the performance of the LS optimal regression fitting curve. In addition, the R^2 estimates of the

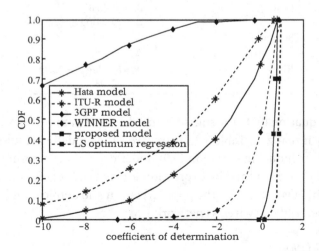

FIGURE 5.10 Comparison of determination coefficients based on the measurement of "Zhengzhou-Xi'an" line of high-speed railway.

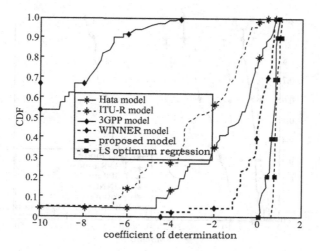

FIGURE 5.11   Comparison of determination coefficients based on the measurement of "Beijing-Shanghai" line of high-speed railway.

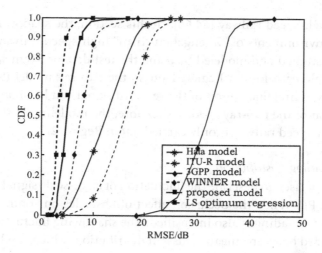

FIGURE 5.12   RMSE comparison based on high-speed railway "Zhengzhou-Xi'an" line measurement.

remaining four types of models in Figures 5.10 and 5.11 appear as negative values, which shows that the predictive performance of the above models is even worse than that of the constant model. Figures 5.12 and 5.13 show the comparison of the RMSE statistical distribution of different models in the Zhengxi line and Beijing-Shanghai line tests, from which it is found that the widely used Hata model and WINNER model have large prediction errors. The RMSE of the model proposed in this paper is very close to the optimal fit and has high prediction accuracy.

5.3.2 Shadow Fading

The research on shadow fading of high-speed railway is the basis of link budget of wireless communication system, which is very important for wireless network design and coverage

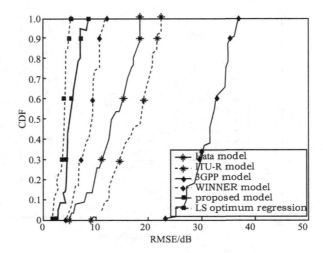

FIGURE 5.13   RMSE comparison based on high-speed railway "Beijing-Shanghai" line measurement.

optimization of high-speed railway [64,66]. In this section, the shadow fading characteristics in various environments of "Zhengzhou-Xi'an" high-speed railway passenger dedicated line will be studied and modeled by using the test data of 53 cm sampling interval. Since the large-scale path loss in Region 1 obeys the two-ray model (Section 5.3.1), the large-scale shadow fading fluctuation of the received signal can be obtained by the deterministic estimation of the two-ray model. Therefore, the modeling of shadow fading and correlation of high-speed railway is only carried out in Region 2.

5.3.2.1 Shadow Fading Distribution

Shadow fading X_{LS} is used to describe the fluctuation of large-scale signals around the local mean of path loss PL. When the occlusion effect of obstacles on communication links is enhanced, the shadow fading is also intensified. The shadowing fluctuation (in decibels) is usually characterized by a zero-mean Gaussian distribution, which can be expressed as:

$$X_{LS} \sim N[0, \sigma_{LS}] \tag{5.11}$$

where σ_{LS} denotes the STD of shadow fading in dB. The measured data show that the distribution of shadow fading in various scenarios of high-speed railway obeys the 0-mean Gaussian distribution in equation (5.11). In addition, we use the Kolmogorov–Smirnov (KS) test with a CI of 95% to validate this assumption. The statistic of the KS test is defined as the maximum value of the absolute difference between the cumulative distribution function (CDF) of the measured shadow fading components Y_m and the CDF of the estimated distribution Y_e, which can be expressed as:

$$D_{KS} = \max\left(\left|\text{CDF}(Y_m) - \text{CDF}(Y_e)\right|\right) \tag{5.12}$$

where CDF(·) denotes the CDF of the target variable; |·| denotes the absolute value of (·). In order to measure the goodness of fit of the statistical distribution, the KS test compares

D_{KS} with a certain threshold value D_{th} under a given confidence level α_{KS}. In the analysis in this section, the confidence level is set to 95%, ($\alpha_{KS} = 5\%$), therefore, the threshold value $D_{th} = 1.36/\sqrt{N}$ [68]. $D_{KS} > D_{th}$ is rejected with 95% significance, whereas any distribution for which $D_{KS} \le D_{th}$ is accepted with the same level of significance. Table 5.4 summarizes the KS test pass rate of 0-mean Gaussian distribution in various environments of high-speed railways. It is found that in generally high-speed railway environments, the pass rate of 0-mean Gaussian distribution KS test is larger than 84%, which shows that the 0-mean Gaussian distribution model in equation (5.11) has high accuracy in describing the shadow fading of high-speed railway environment.

The mean value of the STD of shadowing in each environment is also presented in Table 5.4: It ranges from 2.7 to 3.7 dB. We also note that σ is significantly less in HSR scenarios than in classical cellular systems, e.g., in ref. [25], σ is larger than 7 dB in urban; in ref. [21], $\sigma = 8.2$ dB in suburban; and in ref. [69], σ is larger than 6 dB in urban and rural. This is because, in HSR networks, the high BS leads to clear LOS propagation with reduced shadowing effects.

5.3.2.2 Autocorrelation Characteristics of Shadow Fading

In order to study the autocorrelation characteristics of shadow fading in high-speed railway environment, the following autocorrelation coefficients of shadow fading are defined:

$$\hat{\rho}_{auto}(\Delta d) = \frac{\sum_{i=1}^{N}\left(X_{LS}(d_i) - \overline{X_{LS}(d_l)}\right) \cdot \left(X_{LS}(d_i + \Delta d) - \overline{X_{LS}(d_l + \Delta d)}\right)}{\sqrt{\sum_{i=1}^{N}\left(X_{LS}(d_i) - \overline{X_{LS}(d_l)}\right)^2} \cdot \sqrt{\left(\sum_{i=1}^{N} X_{LS}(X_{LS}d_i + \Delta d) - \overline{X_{LS}(d_l + \Delta d)}\right)^2}} \quad (5.13)$$

where N is the total number of shadow fading samples in a single cell; Δd denotes the distance difference; $\overline{(\cdot)}$ denotes the sample mean. In addition, the measurement of shadow fading of high-speed railway shows that $\hat{\rho}_{auto}$ decreases exponentially with Δd [24]:

$$\hat{\rho}_{auto}(\Delta d) = \exp\left(-\frac{\Delta d}{d_{cor,LS}}\right) \quad (5.14)$$

TABLE 5.4 Environmental Shadow Fading KS Pass Rate, STD, and Autocorrelation Characteristics of HRS

Environment	Pass Rate of KS Test	Mean Standard Deviation/dB	Mean Value of $d_{cor,LS}$/m
Urban	96.97%	3.19	57.12
Suburb	85.48%	3.33	112.48
Rural	93.61%	2.85	114.79
Viaduct	91.92%	2.73	115.44
Cutting	91.60%	3.63	88.78
Station	84.59%	2.77	101.22
River	91.09%	3.09	114.58

where $d_{cor,LS}$ is the shadow fading decorrelation distance, that is, the distance difference when the autocorrelation coefficient $\hat{\rho}_{auto}$ decreases to 1/e. The estimated value of $d_{cor,LS}$ is affected by the measurement environment, which can reflect the change speed of large-scale shadow fading with distance in different environments.

Figure 5.14 shows the measurement example of $\hat{\rho}_{auto}$ in a single cell of the viaduct scenarios. It is found from the figure that the exponential decay model of equation (5.14) is in good agreement with the measurement results. It should be noted that the model range in equation (5.14) is between 0 and 1, while the estimated autocorrelation coefficient in equation (5.13) is between 1 and 1 (negative value $\hat{\rho}_{auto}$ are also found in Figure 5.14). However, measurements show that $\hat{\rho}_{auto}$ is negative only when a is large, and in this region, $|\hat{\rho}_{auto}|$ is less than 0.2, which has little influence on the correlation model. Therefore, the classical exponential decay model in ref. [24] is still used to describe the autocorrelation characteristics of shadow fading (equation 5.14). In addition, the estimation interval of $\hat{\rho}_{auto}$ at 95% confidence is shown in Figure 5.14. It is found that the estimation interval at 95% confidence is very close to the actual estimation of $|\hat{\rho}_{auto}|$ in the region of $|\hat{\rho}_{auto}| > 0.2$, which indicates that the proposed model has high accuracy.

Table 5.4 summarizes the estimated mean values of $d_{cor,LS}$ under multi-cell measurements in seven types of high-speed railway environments, from which it can be found that $d_{cor,LS} > 100\,m$ in most high-speed railway environments; In urban and cutting scenarios, $d_{cor,LS} < 90\,m$, because there are abundant reflection waves in urban and cutting scenarios, which makes the change of statistical characteristics of shadow fading in the environment intensify and $d_{cor,LS}$ decrease. In addition, the measurements in Table 5.4 are also consistent with refs. [4,70], where refs. [4,70] indicate that in an urban LOS environment, $d_{cor,LS} \leq 90\,m$.

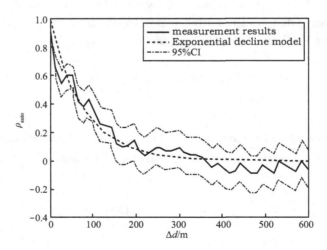

FIGURE 5.14 Example plot of the measured $\hat{\rho}_{auto}$ in one cell of the viaduct environments, together with an exponential decay model.

5.3.2.3 Shadow Fading Cross-Correlation Characteristics of Adjacent Base Stations

In order to study the cross-correlation characteristics of shadow fading of adjacent base stations in high-speed railway environment, the following cross-correlation coefficients of shadow fading of adjacent base stations are defined:

$$\hat{\rho}_{\text{cross}} = \frac{\sum_{i=1}^{N}\left(X_{\text{LS},1,i} - \overline{X_{\text{LS},1}}\right)\cdot\left(X_{\text{LS},2,i} - \overline{X_{\text{LS},2}}\right)}{\sqrt{\sum_{i=1}^{N}\left(X_{\text{LS},1,i} - \overline{X_{\text{LS},1}}\right)^{2}}\cdot\sqrt{\sum_{i=1}^{N}\left(X_{\text{LS},2,i} - \overline{X_{\text{LS},2}}\right)^{2}}} \tag{5.15}$$

where $X_{\text{LS},1,i}$ and $X_{\text{LS},2,i}$ respectively denote the i-th shadow fading sample in two adjacent cells in the same kind of environment. We will analyze and model the estimated values of cross-correlation coefficients in different scenarios based on measurements.

Firstly, the estimation results of cross-correlation coefficients in various scenarios of high-speed railway are analyzed. It is found that the cross-correlation coefficient $\hat{\rho}_{\text{cross}}$ exhibits large fluctuation from cell to cell. It is conjectured that $\hat{\rho}_{\text{cross}}$ is affected by factors such as antennas, BSs, and environments. We thus examine some of these factors in the following and then propose a heuristic model for $\hat{\rho}_{\text{cross}}$. Several factors are considered, i.e., the environment where the neighboring cell is located; the separation distance between the two BSs; the heights h_b of the two BSs; and the tilt angles α_1 of the antennas against the BS towers.

- Environment: It has been found that the environment type significantly affects the large- and small-scale characteristics. Therefore, the study of shadow fading cross-correlation characteristics of adjacent base stations should also distinguish different propagation scenarios. There are enough measurement data for the research and modeling of cross-correlation characteristics in various scenarios.

- Separation distance: Dependence of $\hat{\rho}_{\text{cross}}$ on D is not observed in our measurements. One possible reason is that the separation distance in our measurements is always around 3–4 km, which does not cover a sufficiently large range. We also note that, as reported in ref. [71], a clear dependence of cross correlation can be only observed when the TX/RX separation distance is around 1,000 m, which is not a realistic case for HSR deployment. Since we use the operative GSM-R network in the measurements, changing separation distance to have more realizations is not feasible in our current work. We therefore do not consider the impact of D on $\hat{\rho}_{\text{cross}}$ in the following model.

- h_b and α_1: Under the directional base station antenna system unique to high-speed railway (as shown in Figure 2.4), the base station antenna height h_b and the angle between the antenna and the tower α_1 determine the antenna gain and received signal strength at a specific position in the cell. Larger h_b and smaller α_1 can effectively improve the wireless coverage strength in high-speed railway community.

We therefore introduce a heuristic term of h_b / α_1 to denote the impact of BS on the cross-correlation characteristics. Note that similar parameter combinations such as parameter $h_b / \tan \alpha_1$ are also tested here, but h_b / α_1 has the best fitting performance for the measured results. The adjacent base station difference factor ξ is defined as:

$$\xi = \left| \frac{h_{b,1}}{\alpha_{1,1}} - \frac{h_{b,2}}{\alpha_{1,2}} \right| \tag{5.16}$$

where subscripts 1 and 2 respectively denote two adjacent cells in the same type of high-speed railway environment. A small ξ means that the two neighboring BSs generally have similar impacts on shadow fading of the two TX-RX links, i.e., a large $|\hat{\rho}_{cross}|$ is expected. In Table 5.5, we summarize the number of measured $\hat{\rho}_{cross}$ for each realization of ξ in the measurements. We observe that, except for urban, sufficient realizations enable to examine the dependence of $\hat{\rho}_{cross}$ on ξ, and that for each realization of ξ, we generally have sufficient samples of $\hat{\rho}_{cross}$ to ensure an accurate analysis. For urban environment, the statistical mean of cross-correlation coefficient estimation results is directly given.

Secondly, for the convenience of subsequent elaboration and to describe the statistical modeling method more clearly, the symbolic notation method is defined as follows:

- $\xi_{j,q}$ denotes the q-th realization of ξ in the j-th scenario, where $q = 1, 2, \cdots, Q_j$, and Q_j is the total number of realization ξ in the j-th scenario

- $\phi_{j,q}$ denotes a set of cross-correlation coefficients, which correspond to the NCs with a realization of $\xi_{j,q}$ in the j-th scenario.

- $\hat{\rho}_{cross,j,q,p}$ denotes the p-th cross-correlation coefficient in the set $\phi_{j,q}$, where $p = 1, 2, \cdots, P_{j,q}$, and $P_{j,q}$ is the total number of cross-correlation coefficients in the set $\phi_{j,q}$.

Finally, statistical modeling is performed on the estimates of $\hat{\rho}_{cross}$ in each environment. Figure 5.15 shows examples of estimates of $\hat{\rho}_{cross}$ for viaduct and road cutting environments. It is found from Figure 5.15a and d that the estimated value of a in each set $\phi_{j,q}$ (in each

TABLE 5.5 The Number of Measured $\hat{\rho}_{cross}$ Values for Each Realization of ξ in the Measurements

Environment	Measured Value of ξ	Number of Samples of the Estimated Value of $\hat{\rho}_{cross}$
Urban	(1.6)	(33)
Suburb	(0,0.71,1.25,1.42,1.43,1.71,.96,2.29,2.46,2.86, 3.39,5.71)	(335,18,95,24,26,26,25,78, 26,26,26,25)
Rural	(0,0.40,0.71,1.25,2.14,4.25, 11.79)	(78,107,26,26,26,35,106)
Viaduct	(0,0.15,0.71,1.25,1.75,2.46,2.50,3.00)	(524,16,30,193,103,5,26,106)
Cutting	(0,0.71,1.04,1.25,1.43,1.86, 3.54)	(93,24,26,104,52,20,281)
Station	(0,0.86,3.04,3.54,4.25,5.18, 6.21)	(50,3,14,100,109,26,84)
River	(0,1.5,2.5)	(74,24,26)

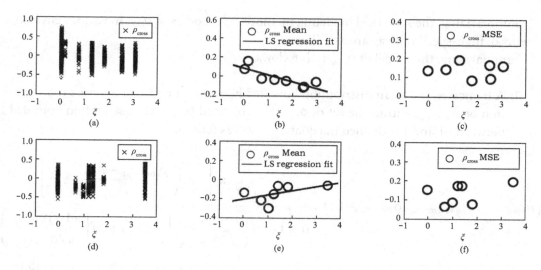

FIGURE 5.15   Example plots of $\hat{\rho}_{cross}$. (a)–(c) Viaduct environment. (d)–(f) Cutting environment.

group $\xi_{j,q}$ and $\hat{\rho}_{cross,j,q,p}$ b) has a larger value between −1 and 1 fluctuation. Furthermore, Figure 5.15b and e show the mean of $\hat{\rho}_{cross,j,q,p}$ within each set $\phi_{j,q}$:

Example plots of ρ_{cross} values measured in the viaduct and cutting environments are shown in Figure 5.7. Our first observation in Figure 5.15a and d is that, within each set of $\phi_{j,q}$ (i.e., for each particular $\xi_{j,q}$), $\hat{\rho}_{cross,j,q,p}$ generally has a large variation, and a distribution ranging between −1 and 1 should be used to describe the variation. Meanwhile, the mean value of $\hat{\rho}_{cross,j,q,p}$ within the set $\phi_{j,q}$, which is defined as:

$$\bar{\rho}_{cross,j,q} = \frac{1}{P_{j,q}} \cdot \left(\sum_{p=1}^{P_{j,q}} \left(\hat{\rho}_{cross,j,q,p} \right) \right) \tag{5.17}$$

is found to follow a linear function of $\xi_{j,q}$, as shown in Figure 15.5b and e. It is observed that in suburban, rural, viaduct, station, and river scenarios, $\bar{\rho}_{cross,j,q}$ decreases with increasing $\xi_{j,q}$, whereas in the cutting scenario, $\bar{\rho}_{cross,j,q}$ is found to increase with $\xi_{j,q}$. Our measurements show that the cutting structure, as discussed in refs. [21] and [37], leads to a negative cross-correlation coefficient at small values of ξ. Note that, in all six scenarios, a small value of $\left| \bar{\rho}_{cross,j,q} \right|$ is generally observed at large $\xi_{j,q}$, which follows the physical insight that a small $\xi_{j,q}$ implies that the scenario difference between the two links is small, and therefore, a large cross correlation is observed. Finally, the STD of $\hat{\rho}_{cross,j,q,p}$ within the set $\phi_{j,q}$, which is defined as:

$$\sigma_{cross,j,q} = \sqrt{ \frac{1}{P_{j,q}} \cdot \left(\sum_{p=1}^{P_{j,q}} \left(\hat{\rho}_{cross,j,q,p} - \bar{\rho}_{cross,j,q} \right)^2 \right) } \tag{5.18}$$

is found to be independent of $\xi_{j,q}$, as shown in Figure 5.15c and f.

Summarizing, the statistical modeling methods and models of $\hat{\rho}_{\text{cross}}$ in various environments of high-speed railway are as follows:

Summarizing, the model of $\hat{\rho}_{\text{cross}}$ is as follows.

1. A truncated Gaussian distribution bounded between −1 and 1 to describe the variation of $\hat{\rho}_{\text{cross},j,q,p}$ within the set of $\phi_{j,q}$. The truncated Gaussian distribution bounded between −1 and 1 is defined in equation (5.19) as follows:

$$f\left(\hat{\rho}_{\text{cross},j,q,p} \in \phi_{j,q}; \overline{\rho}_{\text{cross},j,q}, \sigma_{\text{cross},j,q}, -1,1\right) = \frac{\frac{1}{\sqrt{2\pi}}\exp\left(-0.5\left(\frac{\hat{\rho}_{\text{cross},j,q,p} - \overline{\rho}_{\text{cross},j,q}}{\sigma_{\text{cross},j,q}}\right)^2\right)}{0.5\left[1 + \text{erf}\left(\frac{1 - \rho_{\text{cross},j,q}}{\sqrt{2}\sigma_{\text{cross},j,q}}\right)\right] - 0.5\left[1 + \text{erf}\left(\frac{-1 - \rho_{\text{cross},j,q}}{\sqrt{2}\sigma_{\text{cross},j,q}}\right)\right]}$$

(5.19)

where $\text{erf}(x)$ is the error function:

$$\text{erf}(x) = \frac{2}{\sqrt{\pi}}\int_0^x \exp(-t^2)\,\mathrm{d}t$$

(5.20)

2. $\overline{\rho}_{\text{cross},j,q}$ is modeled as a linear function of $\xi_{j,q}$, which is expressed as

$$\overline{\rho}_{\text{cross},j,q} = a_{\text{cross}} \cdot \xi_{j,q} + b_{\text{cross}}$$

(5.21)

3. a_{cross} and b_{cross} are undetermined coefficients, which can be obtained by LS regression fitting. Note that $\overline{\rho}_{\text{cross},j,q}$ cannot be modeled using equation (5.21) because there is only one set of $\xi_{j,q}$ in urban scenario. Therefore, in the urban environment, let $a_{\text{cross}} = 0$, and use the measured statistical mean of p_{cross} (b_{cross}) to express the cross-correlation coefficient.

In the j-type high-speed railway environment, the overall average of $\sigma_{\text{cross},j,q}$ is used as the model of $\sigma_{\text{cross},j,q}$:

Instead of modeling $\sigma_{\text{cross},j,q}$ as a function of $\xi_{j,q}$, we simply average $\sigma_{\text{cross},j,q}$ in the j-th environment, which is expressed as

$$\sigma_{\text{cross},j,q} = \frac{1}{Q} \cdot \left(\sum_{q=1}^{Q_j}\left(\sigma_{\text{cross},j,q}\right)\right)$$

(5.22)

This is because (1) no distinct dependence of $\sigma_{\text{cross},j,q}$ on $\xi_{j,q}$ is observed, as shown in Figure 5.15c and f, and (2) (17) reduces the estimation error caused by the different $P_{j,q}$ in each set of $\phi_{j,q}$ and avoids misleading conclusions.

In Figure 5.16, we show the example CDF plots of the estimated $\hat{\rho}_{cross}$ in the viaduct and cutting environments. It is found that the truncated Gaussian distribution offers a reasonable fit, which has a KS passing rate larger than 87% in all six environments (the urban environment does not adopt the above statistical modeling method, so it is not considered). The goodness-of-fit of the uniform distribution is also examined, but it generally exhibits a KS passing rate lower than 50%.

Table 5.6 summarizes the parameters of the shadow fading cross-correlation model in the high-speed railway environment and the RMSE of the estimated results of equation (5.21). It is found that the RMSE of the models is generally lower than 0.1, which indicates that the proposed cross-correlation coefficient model has high prediction accuracy. Note that since there is only one set of $\xi_{j,q}$ in the urban environment, $a_{cross} = 0$ in equation (5.21) is required here, so RMSE of b_{cross} is not analyzed in this scenario.

5.3.2.4 Model Validation

Since it is very difficult to conduct practical channel measurements in HSR, a recipe of the generation of the large-scale fading channel is very useful for system design. Based on the

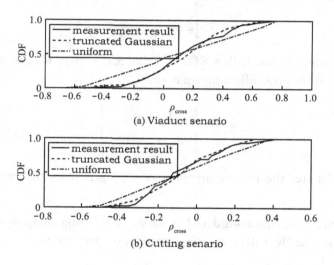

(a) Viaduct senario

(b) Cutting senario

FIGURE 5.16　Example plots of $\hat{\rho}_{cross}$ when $\xi = 0$, together with the CDFs of truncated Gaussian and uniform distributions. (a) Viaduct. (b) Cutting.

TABLE 5.6　Parameters of the Cross-Correlation Model of Shadow Fading in High-Speed Railway Environment

Environment	a_{cross}	b_{cross}	$\bar{\sigma}_{cross,j,q}$	RMSE
Urban	0	0.14	0.20	—
Suburb	−0.055	0.25	0.16	0.08
Rural	−0.016	0.066	0.18	0.07
Viaduct	−0.086	0.16	0.17	0.06
Cutting	0.056	−0.16	0.17	0.09
Station	−0.053	0.23	0.14	0.09
River	−0.016	0.22	0.21	0.03

above modeling results of shadow fading, a series of random sequences satisfying 0-mean Gaussian distribution and containing required correlation characteristics can be generated to simulate shadow fading in high-speed railway environment. The specific steps are as follows [72].

Step 1 Choose a particular environment of HSR.

Step 2 Generate a covariance matrix K using the model of equation (5.14) and Table 5.4. Perform the following factorization: $\mathbf{K} = \mathbf{P}\mathbf{\Lambda}\mathbf{P}^{\mathrm{T}}$, where $\mathbf{P}$ is the matrix whose columns are the eigenvectors of K and Λ is the diagonal matrix of eigenvalues. $(\cdot)^{\mathrm{T}}$ denotes transpose. Generate two independent and identically distributed zero-mean unit-variance Gaussian random variables x_1 and x_2. Sequences S_1 and S_2, both of which have the desired covariance matrix K, can be obtained by

$$\begin{bmatrix} S_1 & S_2 \end{bmatrix} = \left(\mathbf{P}\sqrt{\mathbf{\Lambda}} \right)\begin{bmatrix} x_1 & x_2 \end{bmatrix} \tag{5.23}$$

Step 3 Generate two Gaussian random variables S_1 and S_2 by the following method:

$$\begin{bmatrix} S_1 & S_2 \end{bmatrix} = \sigma_{\mathrm{LS}} \cdot \begin{bmatrix} s_1 & s_2 \end{bmatrix} \cdot \mathbf{R} \tag{5.24}$$

where σ_{LS} is the shadow fading STD in Table 5.4. Matrix $\mathbf{R}$ is an upper triangular matrix that satisfies the following equation:

$$\mathbf{R}^{\mathrm{H}}\mathbf{R} = \begin{bmatrix} 1 & \rho_{\mathrm{cross}} \\ \rho_{\mathrm{cross}} & 1 \end{bmatrix} \tag{5.25}$$

where $(\cdot)^{\mathrm{H}}$ denotes the Hermitian transpose. ρ_{cross} is from the proposed model in Table 5.6.

Step 4 S_1 and S_2 can be thus considered as the shadow fading components of two neighboring BSs in one NC with the desired correlation properties.

To validate the proposed correlation model, we use the measurements from ten cells in another HSR line, i.e., the "Beijing–Shanghai" line, whose measurements were not used in the development of the previous models. The measurement system is the same as reported in Section 5.3.1. We generate the shadow fading components with the same number of samples to the measurements. The generated sequences of shadow fading, with full auto- and cross-correlation properties, are compared with the measurements in the "Beijing–Shanghai" line, and both first- and second-order statistics are validated as follows.

- First-order statistic validation: We compare the CDF of the generated shadow fading components with measurements. Examples of CDF comparisons in suburban and rural environments are presented in Figures 5.17a and 5.18a, where we can see that the generated sequences offer a reasonable fit to the measurements.

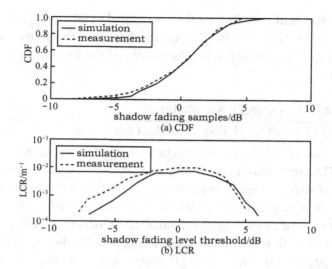

FIGURE 5.17 Validations using the measurements of "Beijing–Shanghai" HSR in a suburban NC, where $\xi = 0$. (a) CDF. (b) LCR.

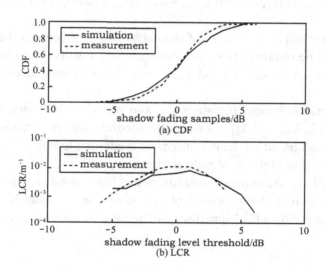

FIGURE 5.18 Validations using the measurements of "Beijing–Shanghai" HSR in a rural NC, where $\xi = 5.83$. (a) CDF. (b) LCR.

- Second-order statistic validation: We use the level crossing rate (LCR) of shadow fading to validate the second-order statistics of our model. The LCR is defined as the number of times that the signal crosses a given threshold level from up to down within a unit of length (1 m) [73]. Examples of LCR comparisons in suburban and rural environments are presented in Figures 5.17b and 5.18b. We find the LCRs of generated shadow fading to be fairly close to the measurements in both cases. Measurements for other environments were also verified, although relevant plots are not shown here due to space limitations.

In this section, the shadow fading model is validated in all high-speed railway environments by using the above methods. The results show that the proposed shadow fading simulation model can accurately predict the change of shadow fading in high-speed railway environment.

5.4 SMALL-SCALE CHANNEL MODELING

5.4.1 Small-Scale Channel Modeling in Viaduct Scenarios

The propagation loss of wireless signal includes three factors: large-scale path loss PL, shadow fading XLS, and small-scale fading XSS. Path loss denotes the process that the propagation loss of wireless signals increases exponentially with distance. Shadow fading indicates that the loss of multipath signal increases due to obstruction in the process of propagation, and most of them show the fluctuation of local mean value of signals in hundreds of times wavelength range; small-scale fading denotes the signal fluctuation caused by multipath propagation in the wavelength range of tens of times. In decibels, the relationship among them can be expressed as [39]:

$$\text{Attenuation (dB)} = \text{PL} + X_{\text{LS}} + X_{\text{SS}} = \text{PL}_0 + 10 n \log d + X_{\text{LS}} + X_{\text{SS}} \tag{5.26}$$

where PL_0 denotes the constant term of the path loss PL, n denotes the path loss index, both of which can be obtained from the measurement results of the least square (Least Square, LS) regression fitting; X_{LS} and X_{SS} denote the large-scale fading and small-scale fading.

The key of wireless channel measurement and modeling is to describe the variation characteristics of PL, X_{LS}, and X_{SS} in a scientific, accurate, and reasonable way. The statistical modeling method based on the measurement is adopted to reflect the most real propagation characteristics of high-speed railway channel.

This section will discuss in detail the influence of high-speed railway viaduct scenarios on the small-scale fading characteristics of the channel, and establish a statistical model based on the measured data [74,75] to describe the small-scale fading characteristics of the channel.

5.4.1.1 Scenarios Analysis of High-Speed Railway Viaduct

Viaduct is one of the most common scenarios in high-speed railway. In order to ensure the smoothness of rails, avoid the introduction of large gradient of rails and ensure the high-speed and stable running of trains, rails are often laid on viaducts of 10~30 m in high-speed railways. In typical high-speed railway lines, the proportion for the viaduct scenarios is often higher than 30%. For example, in the "Beijing-Shanghai" high-speed railway passenger dedicated line, the proportion for the viaduct scenarios is as high as 86.5%. Therefore, the construction of wireless network in viaduct scenarios is very important to ensure the communication security of high-speed railway. Figure 5.19 is a structural diagram of the high-speed railway viaduct scenarios, where H denotes the height of the viaduct. It is found from the figure that there is strong LOS propagation in the viaduct scenarios, and the tall

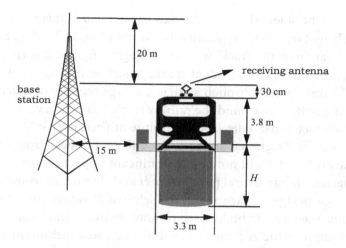

FIGURE 5.19   Structural schematic of high-speed railway viaduct scenarios.

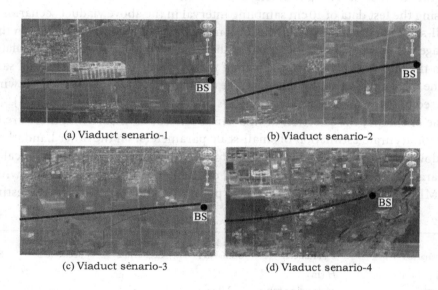

(a) Viaduct senario-1   (b) Viaduct senario-2

(c) Viaduct senario-3   (d) Viaduct senario-4

FIGURE 5.20   Aerial view of various viaduct survey scenarios (*solid points* indicate base station positions and *solid lines* indicate railway tracks).

bridge deck will block some ground backscattered waves. Therefore, the large-scale and small-scale propagation characteristics in the viaduct scenarios will change greatly compared with the traditional public network scenarios.

In order to deeply study the small-scale fading characteristics of the channel in viaduct scenarios, channel measurements were carried out in four types for the viaduct cells. Figure 5.20 shows the Aerial view of four types of viaduct communities. From the figure, it is found that there are a small amount of vegetation and buildings with a height lower than 10 m in all four types of scenarios. In Viaduct Scenario 1, there is an abandoned factory near the track with a height of less than 5 m. There is a small village in the distance with an average height of less than 8 m. In viaduct scenario 2, there is a large open land, and the average height of surface vegetation is less than 8 m. In Viaduct Scenario 3, there is a small village far from the track, with an average height of less than 10 m. Viaduct scenarios 1–3 can be classified as viaduct sparse reflection environment. In viaduct scenario 4, when the train passes through a village, a large number of reflection waves will be generated in the buildings around the railway track. Therefore, viaduct scenario 4 can be classified as viaduct dense reflection environment (because GSM-R adopts the current coverage based on directional antenna, the villages far away from the railway track in viaduct scenarios 1 and 3 will not have a significant impact on wireless propagation). Table 5.7 summarizes the structural parameters H and channel measurement parameters in various scenarios bridge scenarios, and the selected H values cover the typical situation for the viaduct scenarios in high-speed railway environment. In order to increase the sample size and improve the accuracy of analysis, repeated measurements were carried out in four scenarios.

Next, the small-scale fading characteristics for the viaduct scenarios will be deeply studied by using the test data of 10 cm sampling interval in the above viaduct scenarios. Before the small-scale parameter analysis, the local mean value of the measured data in the moving average window of 40X wavelength (i.e., the large-scale sample value) is calculated first, and then the local mean value in the sliding window is divided by the original sample to obtain the local small-scale sample data. Please note that the 40X wavelength window is widely used to extract small-scale parameters of macro honeycomb channel [54].

On the one hand, the selection of this window needs to ensure that there are enough samples for accurate estimation of small-scale parameters; on the other hand, the size of the window needs to be as small as possible to meet the accurate analysis of "local" small-scale characteristics. At a sampling interval of 10 cm, the 40-fold wavelength window in the 930 MHz band can provide 130 sample points for small-scale parameter estimation.

TABLE 5.7   Scenarios Parameters and Model Results of Various Viaducts

Viaduct Scenarios	Parameter	Viaduct Scenario 1	Viaduct Scenario 2	Viaduct Scenario 3	Viaduct Scenario 4
Structural parameters	H (m)	10	15	20	25
Measurement parameters	Measuring range (m)	2,000	2,665	2,502	1,336
	Measurement times	3	2	3	3
	Average speed (km/hour)	260	295	180	65
Results of fade depth (dB)	1%	−14.89	−13.44	−14.28	−14.84
	50%	1.07	0.94	1.14	1.05
	Fading depth	15.96	14.38	15.42	15.89
Rate of the small-scale best fit distribution	Ricean	81.02%	92.29%	95.69%	76.01%
	Nakagami	13.88%	6.42%	3.64%	18.33%
	Rayleigh	4.94%	1.28%	0.67%	5.12%
	Lognormal	0.16%	0%	0	0.27%
	Suzuki	0	0	0	0.27%

At the same time, ref. [54] indicates that the 40X wavelength can be used to analyze the "local" characteristics of channels. Therefore, the extraction of small-scale parameters and the solution of large-scale mean in the following are based on the 40X wavelength window.

5.4.1.2 Fading Depth

Fading depth is used to measure the fluctuation of channel energy relative to local mean caused by small-scale fading, and can be defined as the difference between 50% level and 1% level of small-scale fading signals [76]. Table 5.7 summarizes the fading depth statistics of all test data in four types for viaduct scenarios, from which it can be found that the fading depth for viaduct scenario is around 15 dB, which is smaller compared to the 18.5 dB for the Rayleigh channel (according to the definition of fading depth, the fading depth for the Rayleigh channel can be obtained analytically based on CDF of Rayleigh fading). This is because the antenna at the top of the train in high-speed railway is only slightly higher than the roof (as shown in Figure 5.19), and it is difficult for the backscattered waves from the ground and track surface to reach the receiving antenna through reflection and diffraction. Even in Scenario 4 near the orbit in Reflectors and Scatterers, the 25 m high viaduct still blocks most of the backscattered waves, thus weakening the small-scale deep fading of the channel. In addition, from the comparison in Table 5.7, it can be found that the fading depth of scenario 2 is only 14.38 dB, because the Reflectors and Scatterers in scenario 2 are far away from the track surface. In Scenario 4, the fading depth is increased by 1.5 dB compared with Scenario 2 due to the proximity of the surrounding Reflectors and Scatterers. Therefore, the obstacles around the railway track in the viaduct scenarios still affect the small-scale fading depth of the channel.

5.4.1.3 Distribution Characteristics of Small-Scale Fading

Due to the clear LOS in the high-speed railway viaduct scenarios, the small-scale fading of the channel usually obeys Ricean distribution [11,77]. However, Nakagami, Lognormal, and Suzuki distributions can also be used to describe small-scale fading characteristics in the LOS scenario [78–80]. Therefore, in order to accurately describe the small-scale fading characteristics of high-speed railway viaduct scenarios, it is necessary to adopt appropriate methods to compare the fitting performance of the common small-scale distributions mentioned above, and then determine the optimal small-scale fading distribution function in viaduct scenarios.

The decision method based on Akaike Information Criteria (AIC) is used to select the most suitable small-scale distribution function for high-speed railway viaduct scenarios from Ricean, Rayleigh, Nakagami, Lognormal, and Suzuki. AIC is a measure of the relative goodness of fit of a statistical model and has been widely used in the selection of distribution fit in recent years [81–83]. The AIC for the j-th candidate distribution is given by [81]:

$$\text{AIC}_j = -2\sum_{n=1}^{N} \lg\left[g_{\bar{\theta}_j}(x_n) \right] + 2U \tag{5.27}$$

where $g_{\hat{\theta}_j}$ is the PDF of the j-th distribution function; $\hat{\theta}_j$ is the maximum likelihood estimation of the distribution parameter vector θ_j from the measurements; U is the dimension of vector $\hat{\theta}_j$; N is the dimension size of sample set $x = x_1, x_2, \cdots, x_N$. The model with the lowest AIC provides the best fit. However, the AIC decision is a borderless estimate, and the estimated value of AIC may be very large in some specific cases. To conveniently compare the relative fit of each distribution within the candidate set, we define the AIC differences:

$$\phi_j = \text{AIC}_j - \min_i(\text{AIC}_i) \tag{5.28}$$

where $\min_i(\text{AIC}_i)$ denotes the minimum AIC value over all J candidate families. Then we examine the candidates' relative fitting quality based on the Akaike weights w_j, defined as:

$$w_j = \frac{\exp\left(\dfrac{-\phi_j}{2}\right)}{\displaystyle\sum_{i=1}^{J} \exp\left(\dfrac{-\phi_j}{2}\right)} \tag{5.29}$$

where $\sum_{j=1}^{J} w_j = 1$. The model with the highest Akaike weights is the best distribution to describe the data set.

A sliding and non-overlapping 40X wavelength window is used to estimate the small-scale fading distribution function. Figure 5.21 shows the estimated AIC weight coefficients based on all measured samples in the viaduct scenario.

The following two points can be seen from Figure 5.21.

- Ricean distribution has the best fitting performance, which is consistent with the wide application of Ricean distribution in the LOS scenario.

- In four types for viaduct scenarios, the fitting performance of Lognormal and Suzuki distribution is poor. Note that Suzuki distribution is usually used to describe the fading characteristics of composite channels, that is, channels containing both large-scale and small-scale fading characteristics. The poor fitting performance of Suzuki distribution also proves that the 40X wavelength window can effectively remove the large-scale local mean, and can be used to analyze the small-scale fading characteristics for viaduct scenarios.

Table 5.7 summarizes the optimal fitting ratios of various small-scale distribution functions in four scenarios. It is found from Table 5.7 that the rates of the best fit of Ricean distribution in scenarios 1–3 are 81.02%, 92.29%, and 95.69%, respectively. This is because the larger H value makes the backscattered wave energy of the receiving end decrease. Therefore, the higher the height of the viaduct, the better the fitting characteristics of Ricean distribution. However, a large number of scatterers around the track in scenario 4

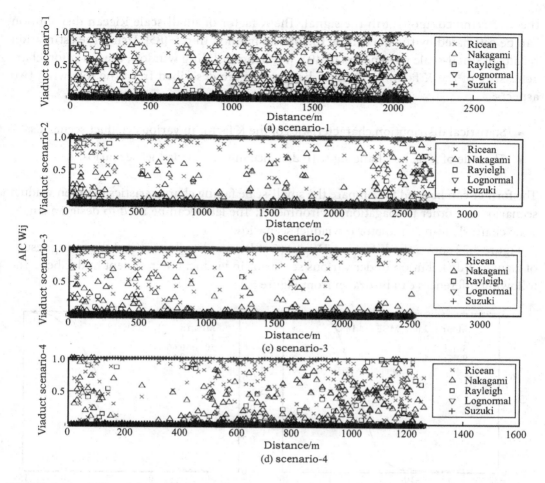

FIGURE 5.21 AIC weight coefficients of various alternative distribution functions in high-speed railway viaduct scenarios.

lead to a drop of best fit rate of Ricean distribution to 76.01%. To sum up, Ricean distribution is used as the small-scale fading distribution of high-speed railway viaduct scenarios.

5.4.1.4 K-Factor Model of Ricean Distribution

The PDF of a Ricean distributed small-scale fading signal can be expressed as [84]:

$$f(r) = \frac{2(K+1)r}{\Omega} \exp\left(-K - \frac{(K+1)r^2}{\Omega}\right) I_0\left(2\sqrt{\frac{K(K+1)}{\Omega}}\, r\right) \qquad (5.30)$$

where r is the amplitude of small-scale fading signal; $I_0(\cdot)$ is a modified Bessel function of order 0 of the first kind; $\Omega = E[r^2]$, and $E[\cdot]$ denote the expectation of the variable. The parameter K is called the Ricean distribution K factor, which is used to express the ratio of the energy of the dominant component (i.e., the direct radius) to the energy of the scattering component in the signal. The K-factor of small-scale Ricean distribution can be estimated by the maximum likelihood method in ref. [85]. Using the estimation results of small-scale parameters in 40X wavelength sliding window, the variation characteristics of the K factor of Ricean distribution are described from the following two aspects:

- Statistical distribution characteristics of the K factor in various viaduct scenarios;

- The rule of K factor changing with distance in various viaduct scenarios.

The former can be used to compare the small-scale fading characteristics between viaduct scenarios and other propagation environments; The latter can be used to design a small-scale channel fading simulator in viaduct scenarios.

Figure 5.22 shows the K-factor CDF curve of Ricean distribution and the fitting results of Gaussian distribution under various scenarios in viaduct environment in decibels. The following inferences can be drawn from Figure 5.22.

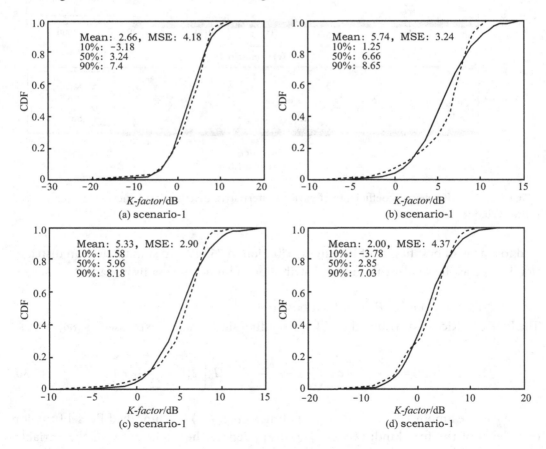

FIGURE 5.22 Ricean K-factors for each scenario. The *dashed* and *solid lines* denote the CDFs of the estimated K-factors from the measurements and the normal fits.

- In Viaduct Scenario 1, the mean K-factor is 2.66 dB, while in Scenario 2 and Scenario 3, the mean K-factor is 5.74 and 5.33 dB, respectively. This means a higher value of H leads to a higher value of the K-factor. This is because most scatterers are lower than the viaduct surface, which results in fewer reflected and scattered multipath components at the receiver.

- Scenario 2 has the highest K factor, with an average of 5.74 dB. This is because Scenario 2 has the least Reflectors and Scatterers in its environment. This shows that the K factor of Ricean distribution is affected by the bridge height and the density of surrounding obstacles.

- Scenario 4 has the lowest K factor, with an average of only 2.0 dB. This is because, in scenario 4, the train passes through a small town, and there are some high buildings located around the track, at a distance of 10–50 m away from it. In such a scenario, the scatterers are higher than the viaduct surface and the reflected and scattered paths relatively easily get to the receiver antenna. Therefore, the severity of fading increases and a low mean value of K is observed. Moreover, the large number of the scatterers in case 4 also results in the highest STD of the K-factor at 4.37 dB. This shows that scatterers that are higher than the viaduct surface and close to the viaduct have a great effect on the severity of fading.

Moreover, the general statistical results of the K-factor derived from the measurements in all four cases are obtained. The mean value and STD of the K-factor over the four scenarios are 4.16 and 3.94 dB.

Next, the variation of the K factor of Ricean distribution with distance is studied. Figure 5.23 shows the variation trend of Ricean distribution K factor with distance in

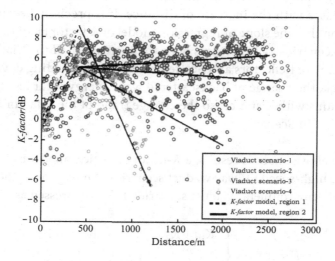

FIGURE 5.23 Measurement results and model of Ricean distribution K factor in high-speed railway viaduct environment.

various scenarios of high-speed railway viaduct environment (the data of each scenario in the figure are statistically averaged on the basis of repeated measurements). It is found from the figure that at the junction of Region 1 and Region 2 under the coverage of the directional antenna system (as shown in Figure 5.4), that is, at the position of 400 m in Figure 5.23, the change trend of the K factor presents a segmented phenomenon. This phenomenon can be interpreted as follows.

- In Region 1, the mobile station is outside the antenna sidelobe coverage area, Figure 5.23, which makes the LOS less powerful at the receiving end. The closer the mobile station is to the base station, the more obvious the attenuation of the LOS power is, and the small-scale fading is aggravated, so the K factor decreases (even lower than 0 dB). As the mobile station gradually moves away from Region 1, the direct radius power increases, so that the K factor increases with the increase of distance within Region 1.

- In Region 2, in Viaduct Scenarios 1–3, the slope of the regression fitting curve of K factor with distance increases with the increase of H. In Scenario 1, the K factor decreases with the increase of distance, because the height of the viaduct in Scenario 1 is low, and there are moderate and stable backscattered waves in the environment, and the energy change of these backscattered waves is much smaller than that of the direct radius with distance (where the intensity of the direct radius is affected by the local shadow effect and decreases with the increase of distance). This phenomenon is consistent with the measurement results in ref. [11]. For Viaduct Scenario 2 and Scenario 3, because the height of the bridge is higher than the Reflectors and Scatterers in the environment, the energy of the backscattered wave is lower and fluctuates greatly, which makes the received backscattered wave energy change faster than the direct radius. This is the reason why the K factor increases with the increase of distance. It is precisely because of the above phenomenon that the slope of regression model of K factor changing with distance in viaduct scenarios 1–3 increases with the increase of H. In Viaduct Scenario 4, because there are Reflectors and Scatterers in the surrounding environment of the railway track, the change of backscattered wave energy is far less than the change of direct radius with distance, so the change trend of the K factor with distance is similar to that in Scenario 1.

According to the above analysis results, a K-factor statistical model based on distance is established in the high-speed railway viaduct scenarios. Among them, the K-factor model in sparse reflection environment (viaduct scenarios 1–3) is expressed as:

$$K_{\mathrm{Mod}}(\mathrm{dB}) = \begin{cases} K_1 d + K_2 & ,\text{Region 1} \\ \left(aH + \dfrac{b}{H} + c\right)d + K_3 & ,\text{Region 2} \end{cases} \qquad (5.31)$$

The K-factor model in dense reflection environment (viaduct scenario 4) is expressed as:

$$K_{\text{Den}}(\text{dB}) = \begin{cases} K_4 d + K_5, \text{Region 1} \\ K_6 d + K_7, \text{Region 2} \end{cases} \tag{5.32}$$

where d is the horizontal distance between the transmitting station and the receiving station; a, b, c, and $K_1 \sim K_7$ are undetermined coefficients.

In viaduct scenarios 1–3, because the variation law of K factor in Region 1 is very similar, the parameter K_1 and K_2 can be obtained by linear regression fitting based on LS. The parameters a, b, and c can be obtained from the slope of the LS fitting function in Region 2:

$$aH + \frac{b}{H} + c = -0.000\,37H - \frac{0.18}{H} + 0.017 \tag{5.33}$$

Equation (5.31) should be a continuous function at the boundary between Regions 1 and 2, so:

$$400K_1 + K_2 = 400\left(aH + \frac{b}{H} + c\right) + K_3 \tag{5.34}$$

Then it is solved:

$$K_3 = 0.148H + \frac{72}{H} - 1.71 \tag{5.35}$$

In Viaduct Scenario 4, the dense backscatter environment, the parameters K_4 and K_5 can also be obtained by LS linear regression fitting. The parameter K_6 should obey the structure similar to equation (5.31). However, the larger H value in Scenario 4 cannot improve the estimation result of K factor, because the surrounding Reflectors and Scatterers are much higher than the bridge deck at this time. Therefore, the scatterer "effective height" factor is introduced to improve the accuracy of statistical fitting. The K factor can then be established as a function of the height difference between the height of the viaduct and the "effective height" of the scatterer. Therefore, K_6 can be expressed as:

$$K_6 = -0.000\,37(H - H') - \frac{0.18}{H - H'} + 0.017 \tag{5.36}$$

$K_6 = -0.019$ can be calculated from the measurements by LS regression fit. Therefore, $H' = 19.71$ can be obtained from equation (5.36). In addition, equation (5.32) is a continuous function, so K_7 can be obtained in a similar way to equation (5.34):

$$K_7 = 0.148H + \frac{72}{H - 19.71} - 0.56 \tag{5.37}$$

Summarizing, the K-factor model of Ricean distribution in viaduct scenarios can be expressed as follows in sparse reflection environment (scenarios 1–3):

$$K_{\text{Mod}}(\text{dB}) = \begin{cases} 0.012d + 0.29 & ,\text{Region 1} \\ \left(-0.000\,37H - \dfrac{0.18}{H} + 0.017\right)d + \left(0.148H + \dfrac{72}{H} - 1.71\right), \text{Region 2} \end{cases} \tag{5.38}$$

In a dense backscatter environment (scenario 4), it can be expressed as:

$$K_{\text{Den}}(\text{dB}) = \begin{cases} 0.025d - 0.84 & ,\text{Region 1} \\ \left(-0.000\,37H - \dfrac{0.18}{H-19.71} + 0.024\right)d + \left(0.148H + \dfrac{72}{H-19.71} - 0.56\right), \text{Region 2} \end{cases}$$
$$\tag{5.39}$$

Equations (5.38) and (5.39) describe in detail the characteristics of small-scale fading with distance in viaduct scenarios. Figure 5.23 shows the comparison between the model and the test data, from which it is found that the estimated value of K factor is in good agreement with the measured results. It should be noted that the K-factor model proposed in this section is different from the measurement results in the existing literature, for example:

$$K(\text{dB}) = \begin{cases} 3.7 + 0.019d, & \text{Rural, 5.25 GHz, ref. } [65] \\ 3.0 + 0.014d, & \text{Urban, 5.8 GHz, ref. } [65] \\ 17.1 - 0.020\,5d, & \text{Suburb, 5.25 GHz, ref. } [2] \\ 5.36 - 1.8\log\left(\dfrac{d}{1\,000}\right), & \text{Suburb, 2.5 GHz, ref. } [74] \end{cases} \tag{5.40}$$

All the above models show that the K factor is a monotone function changing with distance. For example, in ref. [4,104], the K factor decreases monotonously with the increase of distance; In ref. [77], the K factor increases monotonously with the increase of distance. However, the measurement results of high-speed railway viaduct scenarios show that the K factor first increases and then decreases with distance, and the breakpoint is located at the boundary point between Region 1 and Region 2 in Figure 5.4. This is caused by the weakening of direct radius near the base station and the enhancement of fading in the directional base station antenna system of high-speed railway.

5.4.2 Small-Scale Channel Modeling in Cutting Scenario

This section will discuss in detail the influence of high-speed railway cutting characteristic scenarios on the small-scale fading characteristics of the channel, and establish a statistical model [86] based on the joint analysis of measured data and cutting structure parameters to describe the small-scale fading characteristics of the channel under the cutting structure.

5.4.2.1 Cutting Scenarios Analysis of High-Speed Railway

Cutting scenarios widely appear in high-speed railway environment. Similar to the function of viaduct, cutting is mainly used for laying rail surface in uneven areas of the ground to reduce the slope of rail surface. The most common cutting structure is the deep cutting shown in Figure 5.24. The cutting has a central symmetrical structure, with steep inclined walls on both sides, which are often covered with vegetation and concrete. The special deep pit structure of cutting leads to a large number of backscattered waves in cutting. In the actual high-speed railway communication, because the receiving antenna of the train roof is often lower than the upper edge of the cutting, the reflection wave inside the cutting will have a great influence on the received signal. Although the direct radius always exists, the abundant backscattered waves in the cutting structure still greatly aggravate the small-scale fading of the channel. Therefore, the study of propagation loss and fading characteristics in cutting scenarios directly affects the quality of wireless communication network in cutting scenarios of high-speed railway. In this section, firstly, the characteristics of small-scale fading in cutting scenarios are studied.

In order to deeply study the cutting scenarios, the channel measurement was carried out in five types of cutting communities. Figure 5.25 shows the aerial view of five types of cutting communities. In Scenarios 1 and 5, there are scattered buildings with 1~2 floors and a small amount of vegetation outside the cutting. In Scenarios 2–4, there is an open field outside the cutting, and the vegetation height is less than 2 m. The structural parameters of cutting in various scenarios are summarized in Figure 5.24 and Table 5.8 respectively. Among them, the upper edge width w_{up} and the lower edge width w_{down} of all kinds of cutting are obtained based on the average of the measurement results at five different

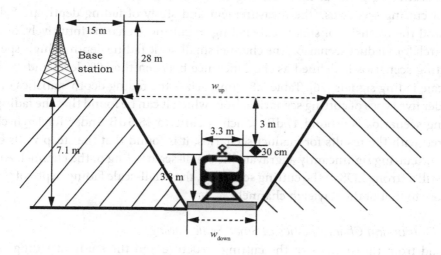

FIGURE 5.24   Section of cutting scenarios of high-speed railway.

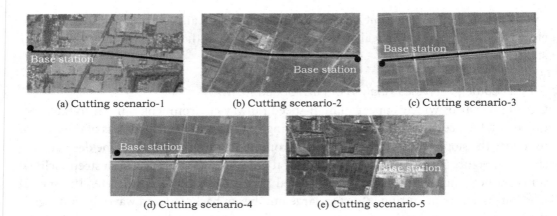

(a) Cutting scenario-1   (b) Cutting scenario-2   (c) Cutting scenario-3

(d) Cutting scenario-4   (e) Cutting scenario-5

FIGURE 5.25 Aerial view of various cutting survey scenarios (*solid points* indicate base station positions, and *thickened solid lines* indicate rail tracks).

positions in the same cutting district. The above five types of cutting scenarios cover the typical characteristics of cutting scenarios in high-speed railway environment. In order to increase the sample size and improve the accuracy of analysis, repeated measurements were carried out in five scenarios.

Next, the small-scale fading characteristics of cutting scenarios will be deeply studied by using the test data with 10 cm sampling interval in the above cutting scenarios. Similar to the previous research on viaduct scenarios, before the small-scale parameter analysis, a moving average window with 40X wavelength is used to remove the large-scale features in the original measurement data (the average value of samples in the sliding window).

5.4.2.2 Fading Depth

Due to the special deep pit structure of cutting, there are a large number of multiple reflection and scattering paths at the receiving end, which greatly aggravates the small-scale fading of cutting scenarios. The measurement and study of fading depth are helpful to understand the intensity of small-scale fading in cutting scenarios intuitively. Similar to the research for viaduct scenarios, the channel small-scale fading depth in high-speed railway cutting scenarios is defined as the difference between the 50% level and 1% level of small-scale fading signals [76]. Table 5.8 summarizes the fading depth statistics of all test data under five types of cutting scenarios, from which it can be found that the fading depth of cutting scenarios is around 17 dB, which is close to 18.5 dB under Rayleigh channel. Compared with the results for viaduct scenarios, it is found that the steep walls on both sides of the cutting significantly aggravate the small-scale fading of the channel. Although there is still a strong LOS in the cutting scenarios, the small-scale fading depth of the channel is close to that of the Rayleigh channel.

5.4.2.3 Distribution Characteristics of Small-Scale Fading

It is found from the analysis of the cutting structure and the study of fading depth in the previous paper that although there is a LOS in the cutting scenarios of high-speed

TABLE 5.8   Scenarios Parameters and Model Results of Various Cutting

Cutting Scenarios	Parameter	Cutting Scenario 1	Cutting Scenario 2	Cutting Scenario 3
Structural parameters	w_{up}/m	58.30	50.86	52.01
	w_{down}/m	15.16	16.85	18.77
Measurement parameters	Measuring range(m)	1,332	1,013	667
	Measurement times	4	3	3
	Average speed (km/hour)	260	290	295
Results of fade depth/dB	1%	−16.19	−17.24	−16.20
	50%	1.23	1.29	1.16
	Fading depth	17.43	18.53	17.37
Rate of the small-scale best fit distribution	Ricean	80.62%	72.34%	78.43%
	Nakagami	17.22%	25.96%	16.99%
	Rayleigh	1.91%	1.28%	3.92%
	Lognormal	0.24%	0.43%	0.65%
	Suzuki	0	0	0
K-factor model parameters	k_1	0.024 5	0.033	0.031 6
	k_2	3.851	1.643	4.519
	k_3	−0.003 82	−0.001	−0.008 11

Cutting Scenarios	Parameter	Cutting Scenario 1	Cutting Scenario 2
Structural parameters	w_{up}/m	55.26	55.72
	w_{down}/m	18.57	18.25
Measurement parameters	Measuring range(m)	689	1,421
	Measurement times	3	3
	Average speed (km/hour)	133	72
Results of fade depth/dB	1%	−15.57	−15.18
	50%	1.20	1.13
	Fading depth	16.79	16.94
Rate of the small-scale best fit distribution	Ricean	77.85%	79.52%
	Nakagami	17.72%	18.37%
	Rayleigh	3.80%	2.11%
	Lognormal	0.63%	0
	Suzuki	0	0
K-factor model parameters	k_1	0.014 3	0.032
	k_2	3.078	2.934
	k_3	−0.003 31	−0.001 64

railway, the existence of steep walls on both sides of the cutting introduces a large number of backscattered waves, which aggravates the small-scale fading of the channel. Under this semi-enclosed structure, the energy specific gravity of direct, reflected, and scattered waves in the received signal is different from that in the traditional propagation scenarios. Therefore, it is necessary to study the distribution characteristics of small-scale fading in cutting scenarios.

Similar to the previous analysis for viaduct scenarios, in the sliding window of the 40X wavelength, the AIC-based decision method is used to select the most suitable small-scale distribution function for high-speed railway cutting scenarios from Ricean, Rayleigh, Nakagami, Lognormal, and Suzuki. Figure 5.26 shows the estimated values of AIC weight

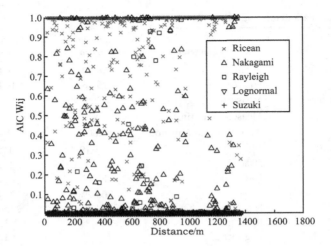

FIGURE 5.26   AIC weight coefficients of various alternative distribution functions in high-speed railway cutting scenario 1.

coefficients based on all measured samples in cutting scenarios 1 (due to space limitation, the estimated results of AIC weight coefficients in other scenarios are no longer given in the form of pictures), among which the AIC weight coefficients of Ricean distribution are the highest. Table 5.8 summarizes the optimal fitting ratios of various small-scale distribution functions under five cutting scenarios. It is found that Ricean distribution has the best fitting performance as a whole, followed by Nakagami distribution, and Rayleigh, Lognormal, and Suzuki distributions are not suitable for cutting scenarios. This is because stable LOS propagation still exists in cutting scenarios, in which case Ricean distribution often has the best performance [37]. In addition, it is found from Table 5.8 that the rate of best fit of Rayleigh distribution in five cutting scenarios is less than 5%, which shows that although there are abundant backscattered waves in cutting scenarios, the intensity of small-scale fading in the channel is still lower than that in the Rayleigh channel due to the existence of direct radius, which is also consistent with the analysis results of fading depth. To sum up, Ricean distribution is used as the small-scale fading distribution of high-speed railway cutting scenarios.

5.4.2.4 K-Factor Model of Ricean Distribution
In order to deeply analyze the small-scale fading characteristics of channel in cutting scenarios, the K-factor of Ricean distribution based on maximum likelihood estimation is studied and statistically modeled. Similar to the previous analysis for viaduct scenarios, the small-scale parameters are estimated in a sliding window of 40X wavelength, and the variation characteristics of K factor of Ricean distribution are described from two aspects: first, the statistical distribution characteristics of K factor in cutting scenarios; The second is the law of K factor changing with distance in various cutting scenarios. The statistical distribution of the K factor can be used to describe the small-scale fading characteristics of cutting scenarios; The distance-based K factor model can be used to achieve more accurate simulation of small-scale fading channels.

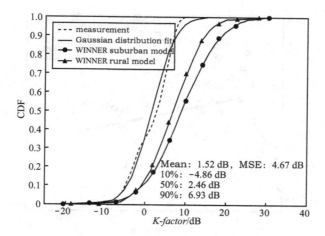

FIGURE 5.27 Statistical distribution of K-factor in 5 cutting scenarios.

Figure 5.27 shows the K-factor CDF curve from the measurements results of five types of cutting scenarios. The CDF of the K-factor is shown in Figure 5.6 and is found to be very nearly Gaussian. The mean value of the K-factor is 1.52 dB and the STD is 4.67 dB. In addition, even though the AIC test rarely gives the Rayleigh distribution as the optimum distribution, the K-factors are often found to be small enough to allow description of the fading as Rayleigh. In contrast, the measurements of the WINNER model are plotted in Figure 5.27 for comparison, which shows that the K-factor in suburban/rural environments is up to 10 dB. The small K-factor in the HSR cutting scenario is caused by the strong reflected and scattered components from the sidewalls.

Next, the variation trend of the K factor with distance in cutting scenarios is analyzed later. Figure 5.28 shows measurements of the K factor in Cutting Scenario 2. It is found from the figure that the K factor increases with increasing distance in Region 1 and decreases with increasing distance in Region 2 in Figure 5.4. This phenomenon is the same as the change trend of K factor in viaduct scenarios, which is mainly caused by directional base station antenna of high-speed railway. Similar to the previous article, the piecewise linear function based on distance is used to statistically model the K factor in cutting scenarios, and the model is expressed as follows:

$$K_{\text{cutting}}(\text{dB}) = \begin{cases} k_1(d-200)+k_2, \text{Region 1} \\ k_3(d-200)+k_2, \text{Region 2} \end{cases} \quad (5.41)$$

where 200 is the segmented point distance obtained based on equation (5.2); k_1, k_2, and k_3 are undetermined coefficients. Table 5.8 summarizes the estimated values of model parameters based on LS linear regression in five types of cutting scenarios. In addition, it can be found that there is a correlation between the model parameters and the cutting structural parameters (w_{up} and w_{down}) in equations (5.41). Therefore, the mathematical relationship between k_1, k_2, and k_3 and the cutting structural parameters w_{up} and w_{down}

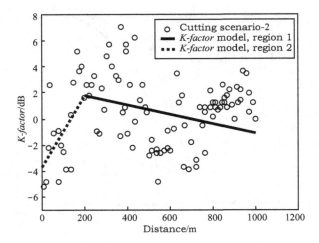

FIGURE 5.28　Measurement results and model of Ricean distribution K factor in cutting 2 scenarios.

is tested in turn by LS regression fitting and fitting accuracy tests, and the methods are as follows.

In the first step, for each target parameter, namely k_1, k_2, and k_3, LS regression fitting is adopted in turn to establish the functional relationship between the target variable and the elements in the following structural parameter set:

$$\text{Sets of Structural Parameters} = \left\{ w_{up}, w_{down}, w_{up} + w_{down}, w_{up} - w_{down}, w_{up} \cdot w_{down}, \frac{w_{up}}{w_{down}} \right\}$$

(5.42)

The elements in the above set cover the most typical forms of cutting structural parameters. Excessive complex structural parameter combinations are not considered here, because the number of actual test scenarios is not enough to support accurate LS regression fitting of complex functions.

In the second step, the Root Mean Square Error (RMSE) and determination coefficient are used to test the accuracy of LS fitting. RMSE and R^2 can be expressed as:

$$\text{RMSE} = \sqrt{\frac{1}{N-1} \sum_{i=1}^{N} (y_i - \bar{y})^2}$$

(5.43)

$$R^2 = 1 - \frac{\sum_{i=1}^{N} (y_i - \hat{y}_i)^2}{\sum_{i=1}^{N} (y_i - \hat{y}_i)^2}$$

(5.44)

where y_i and $\hat{y}_i$ are raw data measurements and model estimates; $\bar{y}$ is the mean value of the measured values of the raw data; N is the total number of samples. RMSE is used to measure the error between the model and the measured sample. R-Square ranges from $-\infty$ to 1, with a value closer to 1 indicating that the regression model fits the data better; and an RMSE value closer to 0 indicates a fit that is more useful for prediction.

Thirdly, in the fitting accuracy judgment of the LS regression model, only the model with the smallest RMSE estimation value and the estimation value larger than 0.5 [59] is selected as the statistical model of the final target parameter. If all the LS regression models do not satisfy the fitting accuracy criteria, it shows that the target parameters are not affected by the structural parameters. Therefore, the target parameters are averaged directly based on the measurement as the final model (constant model) of the target parameters.

Finally, the model fitting accuracy test shows that:

$$\begin{cases} k_1 = 0.027 \\ k_2 = 0.41(w_{up} + w_{down}) - 25.38 \\ k_3 = -0.003\,6 \end{cases} \tag{5.45}$$

The R^2 estimated value of k_2-model is 0.82, which shows that the model can accurately describe the influence of structural parameters on the measurement results. For k_1 and k_3, all LS regression models can't pass the model fitting test, so the constant model based on the average of measurement results is finally adopted.

Summarizing, the K-factor model of Ricean distribution under cutting scenarios can be expressed as:

$$K_{cutting}(dB) = \begin{cases} 0.027d + 0.41(w_{up} + w_{down}) - 30.78, \text{Region 1} \\ -0.003\,6d + 0.41(w_{up} + w_{down}) - 24.66, \text{Region 2} \end{cases} \tag{5.46}$$

Figure 5.28 shows the comparison between the model of equation (5.46) in cutting scenarios 2 and the measurement. It is found from the figure that the proposed model effectively describes the changing trend of K factor in cutting scenarios. In addition, equations (5.46) show that a larger z will cause an increase in the estimated value of K factor, which shows that a wide cutting (the upper and lower edges of the cutting are wide) can effectively reduce the small-scale fading effect of the channel.

5.4.3 Fading Mechanism Research and Channel Modeling Based on Multiple Scattering

In high-speed mobile environment, there are ideal LOS propagation conditions, and Ricean distribution and Nakagami distribution are often used to describe small-scale fading in this environment [87–91]. However, recent studies have shown that Ricean distribution and Nakagami distribution cannot fully reflect the small-scale fading characteristics of wireless channels in all high-speed mobile scenarios [92,93]. The fading mechanism of wireless channel in high-speed mobile scenarios is easily affected by the complex propagation environment. In order to further describe the small-scale fading model in high-speed moving scenarios, multipath signals can be modeled as the sum of multiple scattering energies, and then Second-Order Scattering Fading (SOSF) [94,95] is introduced, which is embodied as follows:

$$S_{\text{SOSF}}(t) = \omega_0 e^{j\theta} + \omega_1 G_1(t) + \omega_2 G_2(t) G_3(t) \tag{5.47}$$

where $\omega_0 e^{j\theta}$ is the LOS component whose weight coefficient is ω_0, and θ is the phase uniformly distributed on $[0, 2\pi]$. G_i is an independent and identically distributed compound normal random variable. The weight coefficient $\{\omega_n\}_{n=0}^{2}$ is a non-negative real number and satisfies that $\omega_0^2 + \omega_1^2 + \omega_2^2 = 1$ denotes the proportion of the LOS component, first-order reflection component and second-order scattering component in the total fading channel, respectively. The cumulative probability distribution function (CDF) for the small-scale fading amplitude $r = |S_{\text{SOSF}}|$ in the SOSF model can be expressed as [94]:

$$F_{\text{SOSF}}(r) = r \int_0^\infty e^{-w_1^2 \kappa^2 / 4} \frac{4 J_1(r\kappa) J_0 w_0 \kappa}{4 + w_2^2 \kappa^2} d\kappa \tag{5.48}$$

where J_0 and J_1 denote the zero-order and first-order Bessel functions, respectively. The SOSF model can be quantified by two parameters [96]:

$$\alpha = \frac{\omega_2^2}{\omega_0^2 + \omega_1^2 + \omega_2^2} \tag{5.49}$$

$$\beta = \frac{\omega_0^2}{\omega_0^2 + \omega_1^2 + \omega_2^2} \tag{5.50}$$

where $\alpha \geq 0, \beta \geq 0\ \alpha + \beta \leq 1$. Obviously, the larger the α value, the larger the ratio of the second-order scattering component to the total fading, which indicates that the fading environment is more complex and the mobile station is in a deep fading region in this region. The above model has a wide range of applications: if $\alpha = 0$ and $\beta > 0$, SOSF distribution becomes Ricean distribution; If $0 < \alpha < 1$ and $\beta \approx 0$, SOSF distribution is approximately Rayleigh distribution plus Double-Rayleigh distribution, abbreviated as RDR distribution; If $0 < \alpha < 1, 0 < \beta < 1$, and $\alpha + \beta < 1$, SOSF distribution becomes a mixture of Ricean distribution and double Rayleigh distribution, which is called the RiDR distribution for short. In addition, the K factor of Ricean distribution can also be obtained from the SOSF distribution parameter [96] $(K = \beta / 1 - \beta)$. It should be pointed out that in the statistical channel model, double Rayleigh distribution is mostly used to describe deep fading under non-LOS propagation conditions, which will lead to an increase in symbol error probability and a decrease in diversity gain [97,98]. In the cutting scenarios of high-speed railway, when the

LOS propagation condition is blocked, the multiple reflection components brought by the car body and the environment, such as the bridge, may lead to double Rayleigh fading.

In the application of this model, the moment estimation method can be used to estimate the weight factor [94]. The moment estimation method assumes $\omega_0 = 0$, that is, there is no LOS propagation condition in the studied scenario, and the remaining weight factors can be estimated by equations (5.51) and (5.52):

$$\hat{w}_1^2 = M_2 - \hat{w}_2^2 \tag{5.51}$$

$$\hat{w}_2^2 = \sqrt{\frac{1}{2}M_4 - M_2^2} \tag{5.52}$$

where M_k denotes the k-th moment of small-scale fading samples. The whole estimation process is estimated according to the maximum likelihood criterion.

Based on the SOSF model, the characteristic analysis of high-speed railway fading channel can be carried out more effectively. Figure 5.29 shows the fitting results and comparison between SOSF distribution of small-scale fading and measured data in high-speed railway cutting scenarios, in which the transmitter is labeled as the transmitter, the receiver is labeled as the receiver and the bridge is labeled as CBr. Figure 5.29 also shows the schematic test scenario (including cutting scenario and crossing bridge), the variation trend of parameter estimation results (α, β) with transceiver distance, and the corresponding Ricean distribution K factor. As can be seen in Figure 5.29, LOS components dominate, but the deep fading effect of double Rayleigh due to non-LOS conditions caused by crossing the

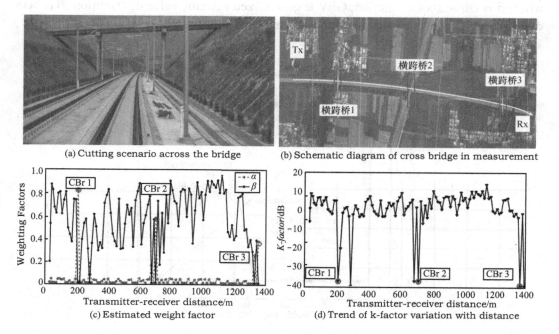

(a) Cutting scenario across the bridge   (b) Schematic diagram of cross bridge in measurement

(c) Estimated weight factor   (d) Trend of k-factor variation with distance

FIGURE 5.29   Estimation results in cutting scenario.

bridge cannot be ignored. The *K*-factor of Ricean distribution is used to describe the characteristics of small-scale fading, and the change trend of the *K*-factor in Figure 5.29 also reflects the influence of crossing the bridge. Similarly, in ref. [99], it is also verified that the *K* factor will fluctuate numerically with the appearance of the span bridge. In fact, crossing the bridge will bring extra path loss and fading. Because LOS propagation conditions often appear in cutting scenarios, it is found that the *K* factor in the above test scenarios obeys exponential distribution, and its value is around 2.1 dB, which is similar to the conclusion of previous research results [100], thus indirectly verifying the rationality of SOSF model in Ricean distribution estimation.

In the above research environment, the estimation results of all SOSF model parameterization (α,β) are shown in Figure 5.30. It is found from the figure that there will be three fading distributions when the SOSF model is applied to high-speed railway cutting scenarios. The Ricean distribution with the LOS propagation component has the highest occurrence probability, followed by RiDR with LOS propagation and double Rayleigh fading, and the RDR distribution has the lowest occurrence probability. It is found that the probability of LOS propagation conditions in cutting scenarios is over 90%, which is consistent with the previous research conclusions. In order to realize the simulation of SOSF model, the probability distribution of (α,β) is obtained by calculating the statistical distribution as follows:

$$(\alpha,\beta) \sim 0.57\delta(\alpha)\text{GEV}(\beta\,|-0.57,0.28,0.47) + 0.08B(\alpha\,|\,0.44,0.53)\delta(\beta)$$

$$+ 0.35N_{\mu=[0.03,0.64],\sigma=[0.01,0.21],\rho=0.99}(\alpha,\beta)$$

(5.53)

where δ is Dirac shock function, GEV is generalized extreme value distribution, B is beta distribution, and N is two-dimensional normal distribution with mean value μ, STD σ, and correlation coefficient ρ.

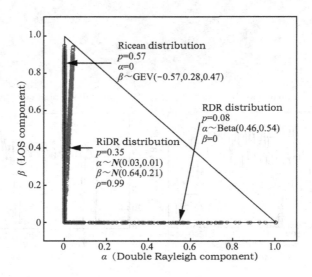

FIGURE 5.30   Estimation results of SOSF distribution parameters based on measured data.

In the fast time-varying high-speed railway channel environment, the distribution process of small-scale fading needs to be described by the implicit Markov process [101] to describe the state transition process of different small-scale fading distributions. Since there are three different fading combination distributions in the SOSF model, a three-state Markov process can be used to describe the state transition probability, as shown in Figure 5.31. Firstly, the (α,β) values of each local interval (within 20 wavelengths) are estimated, three fading distribution states are determined, and the transition probability of fading distribution of each interval and the next interval is calculated. It is found from Figure 5.31 that the transition probabilities from Ricean distribution to RiDR and RDR distribution are 0.34 and 0.09, respectively. The research shows that although Ricean distribution can be used to describe small-scale fading in high-speed rail scenarios, the deep fading distribution represented by double Rayleigh distribution cannot be ignored.

In order to verify the SOSF small-scale fading model in high-speed railway environment, the measured data are used to verify the model, and the fading amplitude of the SOSF model generated by model simulation is compared with the small-scale fading amplitude in the test data. The steps of generating the model are described in refs. [102,103].

Step 1: Generates parameters (α,β) according to the above-mentioned Markov process and three kinds of fading statistical distributions.

Step 2: Calculates the weight factor $\{\omega_n\}_{n=0}^2$ according to (α,β), and then calculates the channel complex random number according to equation (5.47).

Step 3: Running the Markov state transition matrix before calculating the random number of the channel every time, if the fading distribution state is transferred, generating new parameters (α,β), otherwise continuing to carry out channel simulation according to the parameters generated last time.

Figure 5.32 shows a comparison of cumulative probability distribution functions in two fading windows in a high-speed railway environment, using data from non-deep

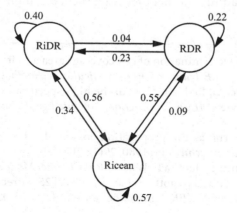

FIGURE 5.31   State transition probability.

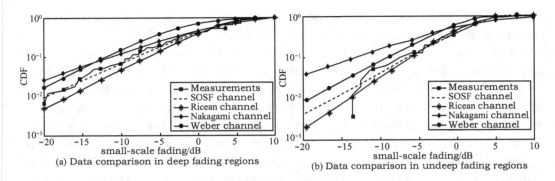

(a) Data comparison in deep fading regions

(b) Data comparison in undeep fading regions

FIGURE 5.32 SOSF model verification based on measured data in high-speed railway scenarios.

TABLE 5.9 Comparison of KS Calibration Results Based on Measured Data

	SOSF	Rice	Nakagami	Weibull
Rejection ratio	19.1%	26.7%	77.1%	98.1%

fading areas and data from deep fading (across bridge occlusion) areas, respectively. From Figure 5.32, it can be found that the SOSF model has a better fitting effect in deep fading region than other reference standard models such as Ricean model, Nakagami model, and Weibull model. In order to further verify the fitting degree between the SOSF model and measured data, Kolmogorov-Smirnov check (KS check) [67] with a significant level of 0.05 is used to verify the measured data with several small-scale fading models mentioned above. According to the KS check results in Table 5.9, the SOSF model can better fit the measured data than other standard fading models in a deep fading region. Therefore, the SOSF model can be applied to describe the small-scale fading characteristics in high-speed railway scenarios. The SOSF model depicts the small-scale fading process of channel under deep fading conditions from the idea of multi-scatterer modeling, which improves the accuracy of small-scale fading channel modeling in high-speed railway scenarios and enriches the small-scale fading channel modeling theory of high-speed railway.

REFERENCES

[1] Ai B, He R, Li G, et al. Determination of cell coverage area and its applications in high-speed railway environments. *IEEE Transactions on Vehicular Technology*, 2017, 66(5): 3515.3525.

[2] He R, Zhong Z, Ai B, et al. Reducing cost of the high-speed railway communications: From propagation channel view. *IEEE Transactions on Intelligent Transportation Systems*, 2015, 16(4): 2050–2060.

[3] Hata M. Empirical formula for propagation loss in land mobile radio services. *IEEE Transactions on Vehicular Technology*, 1980, 29(3): 317–325.

[4] Meinila J, Kyosti P, Jamsa T, Hentilä L. *WINNER II Channel Models*. USA: Wiley, 2009: 39–92.

[5] Molisch A F, Asplund H, Heddergott R, et al. The COST259 directional channel model-part: Overview and methodology. *IEEE Transactions on Wireless Communications*, 2006, 2006(12): 3421–3433.

[6] Ai B, He R, Zhong Z, et al. Radio wave propagation scenarios partitioning for high-speed rails. *International Journal of Antennas and Propagation*, 2012, 2012(1): 1–7.

[7] Lehner A, Garcia C R, Strang T, et al. Measurement and analysis of the direct train to train propagation channel in the 70 cm UHF-band. *Communication Technologies for Vehicles*, 2011, 41(3): 45.57.

[8] Prasad M, Ratnamla K, Dalela P. Mobile communication measurements along railroads and model evaluations over eastern-Indian rural regions. *IEEE Antennas and Propagation Magazine*, 2010, 52(5): 131–141.

[9] Goller M, Masur K, Frohlingsdorf G, et al. Measurement results and parameters for modelling mobile railway radio channels in the 900 MHz band. *Nachrichtentechnik Elektronik*, 1993, 43(6): 290–295.

[10] Ai B, Cheng X, Kurner T, et al. Challenges toward wireless communications for high-speed railway. *IEEE Transactions on Intelligent Transportation Systems*, 2014, 15(4): 1–16.

[11] Greenstein L J, Ghassemzadeh S S, Erceg V, et al. Ricean k-factors in narrow-band fixed wireless channels: Theory, experiments, and statistical models. *IEEE Transactions on Vehicular Technology*, 2009, 58(8): 4000–4012.

[12] Bultitude R J. Measured characteristics of 800/900 MHz fading radio channels with high angle propagation through moderately dense foliage. *IEEE Journal on Selected Areas in Communications*, 1987, 5(2): 116–127.

[13] Stewart K A, Labedz G P, Sohrabi K. Wideband channel measurements at 900 MHz. *IEEE VTC, July 25.28, 1995, Chicago, USA*. Piscataway, NJ: IEEE Press, 1995: 236–240.

[14] Suikkanen E, Hentila L, Meinila J. Wideband radio channel measurements around 800 MHz in outdoor to indoor and urban macro scenarios. *IEEE Future Network and Mobile Summit, June 16–18, 2010, Florence, Italy*. Piscataway, NJ: IEEE Press, 2010: 1–9.

[15] Gao L, Zhong Z, Ai B, Xiong L. Estimation of the Ricean k-factor in the high speed railway scenarios. *IEEE ChinaCom, August 25.27, 2010, Beijing, China*. Piscataway, NJ: IEEE Press, 2010: 1–5.

[16] Gao L, Zhong Z, Ai B, Xiong L, Zhang H. Analysis and emulation of the small-scale fading characteristics in the high-speed rail scenarios. *IEEE ChinaCom, August 17–19, 2011, Harbin, China*. Piscataway, NJ: IEEE Press, 2011: 1181–1185.

[17] Lu J, Zhu G, Briso-Rodriguez C. Fading characteristics in the railway terrain cuttings. *IEEE VTC, May 15.18, 2011, Yokohama, Japan*. Piscataway, NJ: IEEE Press, 2011: 1–5.

[18] Liu L, Tao C, Qiu J, Zhou T, Sun R, Chen H. The dynamic evolution of multipath components in high-speed railway in viaduct scenarios: From the birth-death process point of view. *IEEE PIMRC, September 9–12, 2012, Sydney, Australia*. Piscataway, NJ: IEEE Press, 2012: 1774–1778.

[19] Liu L, Tao C, Qiu J, et al. Position-based modeling for wireless channel on high-speed railway under a viaduct at 2.35 GHz. *IEEE Journal on Selected Areas in Communications*, 2012, 30(4): 834–845.

[20] Herring K T, Holloway J W, Staelin D H, et al. Path-loss characteristics of urban wireless channels. *IEEE Transactions on Antennas and Propagation*, 2010, 58(1): 171–177.

[21] Erceg V, Greenstein L J, Tjandra S Y, et al. An empirically based path loss model for wireless channels in suburban environments. *IEEE Journal on Selected Areas in Communications*, 1999, 17(7): 1205.1211.

[22] Molisch A F. Ultrawideband propagation channels-theory, measurement, and modeling. *IEEE Transactions on Vehicular Technology*, 2005, 54(5): 1528–1545.

[23] Molisch A F, Cassioli D, Chong C C, et al. A comprehensive standardized model for ultra-wideband propagation channels. *IEEE Transactions on Antennas and Propagation*, 2006, 54(11): 3151–3166.

[24] Gudmundson M. Correlation model for shadow fading in mobile radio systems. *Electronics Letters*, 1991, 27(23): 2145.2146.

[25] Algans A, Pedersen K I, Mogensen P E. Experimental analysis of the joint statistical properties of azimuth spread, delay spread, and shadow fading. *IEEE Journal on Selected Areas in Communications*, 2002, 20(3): 523–531.

[26] Sorensen T. Slow fading cross-correlation against azimuth separation of base stations. *Electronics Letters*, 1999, 35(2): 127–129.

[27] Weitzen J, Lowe T J. Measurement of angular and distance correlation properties of lognormal shadowing at 1900 MHz and its application to design of PCS systems. *IEEE Transactions on Vehicular Technology*, 2002, 51(2): 265.273.

[28] Szyszkowicz S S, Yanikomeroglu H, Thompson J S. On the feasibility of wireless shadowing correlation models. *IEEE Transactions on Vehicular Technology*, 2010, 59(9): 4222–4236.

[29] Wei H, Zhong Z, Guan K, Ai B et al. Path loss models in viaduct and plain scenarios of the high-speed railway. *Proceedings of IEEE ChinaCom, August 25.27, 2010, Beijing, China*. Piscataway, NJ: IEEE Press, 2010: 1–5.

[30] Wei H, Zhong Z, Xiong L, et al. Study on the shadow fading characteristic in viaduct scenario of the high-speed railway. *IEEE ChinaCom, August 17–19, 2006, Harbin, China*. Piscataway, NJ: IEEE Press, 2006: 1216–1220.

[31] Guo Y, Zhang J, Zhang C, et al. Correlation analysis of high-speed railway channel parameters based on channel measurement. *IEEE HMWC, November 1–3, 2013, Shanghai, China*. Piscataway, NJ: IEEE Press, 2013: 132–136.

[32] Guo Y, Zhang J, Tao C, et al. Propagation characteristics of wideband high-speed railway channel in viaduct scenario at 2.35 GHz. *Journal of Modern Transportation*, 2012, 20(4): 206–212.

[33] Zhao M, Wu M, Sun Y, et al. Analysis and modeling of the LTE broadband channel for train-ground communications on high-speed railway. *IEEE VTC, September 2–5, 2013, Las Vegas, USA*. Piscataway, NJ: IEEE Press, 2013: 1–5.

[34] Sun R, Tao C, Liu L, et al. Channel measurement and characterization for HSR U-shape groove scenarios at 2.35 GHz. *IEEE VTC, September 2–5, 2013, Las Vegas, USA*. Piscataway, NJ: IEEE Press, 2013: 1–5.

[35] Garcia C R, Lehner A, Strang T, et al. Channel model for train to train communication using the 400 MHz band. *IEEE VTC, May 11–14, 2008, Singapore*. Piscataway, NJ: IEEE Press, 2008: 3082–3086.

[36] Guan K, Zhong Z, Ai B, et al. Empirical models for extra propagation loss of train stations on high-speed railway. *IEEE Transactions on Antennas and Propagation*, 2014, 62(3): 1395.1408.

[37] Molisch A F. *Wireless Communications*. 2nd ed. USA: Wiley, 2010: 102–235.

[38] Mecklenbrauker C F, Molisch A F, Karedal J, et al. Vehicular channel characterization and its implications for wireless system design and performance. *Proceedings of the IEEE*, 2011, 99(7): 1189–1212.

[39] Rappaport T S. *Wireless Communications: Principles and Practice*. USA: Prentice Hall, 1996: 69–138.

[40] Stuber G L. *Principles of Mobile Communication*. USA: Springer, 2011: 608–663.

[41] Tian L, Li J, Huang Y, et al. Seamless dual-link handover scheme in broadband wireless communication systems for high-speed rail. *IEEE Journal on Selected Areas in Communications*, 2012, 30(4): 708–718.

[42] Graziosi F, Pratesi M, Ruggieri M, et al. A multicell model of handover initiation in mobile cellular networks. *IEEE Transactions on Vehicular Technology*, 1999, 48(3): 802–814.

[43] Lin H P, Juang R T, Lin D B. Validation of an improved location-based handover algorithm using GSM measurement data. *IEEE Transactions on Mobile Computing*, 2005, 4(5): 530–536.

[44] He R, Zhong Z, Ai B. Path loss measurements and analysis for high-speed railway viaduct scenarios. *The 6th International Wireless Communications and Mobile Computing Conference (IWCMC), France*. New York: ACM Press, 2010: 266–270.

[45] ITU-R. *Requirements for the Radio Interface(s) for International Mobile Telecommunications-2000: M.1034-1. 39-2000.* Geneva: International Telecommunication Union Radiocommunication Sector, 2000: 4–10.

[46] Mikoshiba K, Nurita Y. Guided radiation by coaxial cable for train wireless systems in tunnels. *IEEE Transactions on Vehicular Technology,* 1969, 18(2): 66–69.

[47] Heddebaut M. Leaky waveguide for train-to-wayside communication-based train control. *IEEE Transactions on Vehicular Technology,* 2009, 58(3): 1068–1076.

[48] Seidel D B, Wait J R. Transmission modes in a braided coaxial cable and coupling to a tunnel environment. *IEEE Transactions on Microwave Theory and Techniques,* 1978, 26(7): 494–499.

[49] Delogne P P, Deryck L. Underground use of a coaxial cable with leaky sections. *IEEE Transactions on Antennas and Propagation,* 1980, 28(6): 875.883.

[50] Lienard M, Degauque P. Propagation in wide tunnels at 2 GHz: A statistical analysis. *IEEE Transactions on Vehicular Technology,* 1998, 47(4): 1322–1328.

[51] Aikio P, Gruber R, Vainikainen P. Wideband radio channel measurements for train tunnels. *IEEE VTC, May 18–21, 1998, Ottawa, Canada.* Piscataway, NJ: IEEE Press, 1998: 460–464.

[52] Lienard M, Betrencourt S, Degauque P. Theoretical and experimental approach of the propagation at 2.5 GHz and 10 GHz in straight and curved tunnels. *IEEE VTC, September 19–22, 1999, Amsterdam, Netherland.* Volume 4. Piscataway, NJ: IEEE Press, 1999: 2268–2271.

[53] Ai B, Zhong Z D, Zhu G, Lin S. Theoretical analysis on local mean signal power for wireless field strength coverage. *IEEE WCSP, November 13–15, Nanjing, China.* Piscataway, NJ: IEEE Press, 2009: 1–5.

[54] Wong D, Cox D C. Estimating local mean signal power level in a Rayleigh fading environment. *IEEE Transactions on Vehicular Technology,* 1999, 48(3): 956–959.

[55] He R, Zhong Z, Ai B, Ding J, Jiang W, Zhang H, Li X. A standardized path loss model for the GSM-railway based high-speed railway communication systems. *IEEE Vehicular Technology Conference (VTC)-Spring, May 18–21, 2014, Seoul, South Korea.* Piscataway, NJ: IEEE Press, 2014: 1–5.

[56] Steinbauer M, Molisch A F, Bonek E. The double-directional radio channel. *IEEE Antennas and Propagation Magazine,* 2001, 43(4): 51–63.

[57] Friis H T. A note on a simple transmission formula. *Proceedings of the IRE,* 1946, 34(5): 254–256.

[58] Rawlings J, Pantula S, Dickey D. *Applied Regression Analysis: A Research Tool.* New York: Springer Verlag, 1998: 205.231.

[59] ITU-R. *Propagation by Diffraction: P.526-12. 02-2012.* Geneva: International Telecommunication Union Radiocommunication Sector, 2012: 15.32.

[60] He R, Zhong Z, Ai B, et al. An empirical path loss model and fading analysis for high-speed railway viaduct scenarios. *IEEE Antennas and Wireless Propagation Letters,* 2011, 10(PP): 808–812.

[61] He R, Ai B, Zhong Z, et al. A measurement-based stochastic model for high-speed railway channels. *IEEE Transactions on Intelligent Transportation Systems,* 2015, 16(3): 1120–1135.

[62] He R, Zhong Z, Ai B, Zhang B. Measurement-based auto-correlation model of shadow fading for the high-speed railways in urban, suburban, and rural environments. *IEEE International Symposium on Antennas and Propagation, July 6–11, 2014, Memphis, USA.* Piscataway, NJ: IEEE Press, 2014: 949–950.

[63] Hu M, Zhong Z, Ni M, et al. Analysis of link lifetime with auto-correlated shadowing in high-speed railway networks. *IEEE Communication Letters,* 2015, 19(12): 2106–2109.

[64] He R, Zhong Z, Ai B, et al. Shadow fading correlation in high-speed railway environments. *IEEE Transactions on Vehicular Technology,* 2015, 64(7): 2762–2772.

[65] He R, Zhong Z, Ai B, Oestges C. A heuristic cross-correlation model of shadow fading in high-speed railway environments. *URSI General Assembly and Scientific Symposium, August 16–23, 2014, Beijing, China.* Piscataway, NJ: IEEE Press, 2014: 1–4.

[66] Lilliefors H W. On the Kolmogorov-Smirnov test for normality with mean and variance unknown. *Journal of the American Statistical Association*, 1967, 62(318): 399–402.

[67] Massey Jr F J. The Kolmogorov-Smirnov test for goodness of fit. *Journal of American Statistical Association*, 1951, 46(253): 68–78.

[68] Fischer J, Grossmann M, Felber W, Landmann M, Heuberger A. Modeling shadow fading for mobile-to-mobile communications in the VHF and UHF band. *IEEE ICWITS, November 11–16, 2012, Hawaii, USA*. Piscataway, NJ: IEEE Press, 2012: 1–4.

[69] Zhu M, Tufvesson F, Medbo J. Correlation properties of large scale parameters from 2.66 GHz multi-site macro cell measurements. *IEEE VTC, May 15.18, 2011, Yokohama, Japan*. Piscataway, NJ: IEEE Press, 2011: 1–5.

[70] Chu Q H, Conrat J M, Cousin J. Experimental characterization and modeling of shadow fading correlation for relaying systems. *IEEE VTC, September 5.8, 2011, San Francisco, USA*. Piscataway, NJ: IEEE Press, 2011: 1–5.

[71] Leon-Garcia A. *Probability, Statistics, and Random Processes for Electrical Engineering*. USA: Pearson Prentice Hall, 2008: 303–358.

[72] Patzold M, Yang K. An exact solution for the level-crossing rate of shadow fading processes modelled by using the sum-of-sinusoids principle. *Wireless Personal Communication*, 2010, 52(1): 57–68.

[73] He R, Zhong Z, Ai B, Ding J. Measurements and analysis of short-term fading behavior for high-speed rail viaduct scenario. *IEEE International Conference on Communication (ICC), June 10–15, 2012, Canada*. Piscataway, NJ: IEEE Press, 2012: 4563–4567.

[74] He R, Zhong Z, Ai B, Ding J. Distance-dependent model of Ricean K-factors in high-speed rail viaduct channel. *2012 IEEE Vehicular Technology Conference (VTC Fall), September 3–6, 2012, Quebec City, QC, Canada*. Piscataway, NJ: IEEE Press, 2012: 1–5.

[75] Lee W C. Estimate of local average power of a mobile radio signal. *IEEE Transactions on Vehicular Technology*, 1985, 34(1): 22–27.

[76] Kozono S. Received signal-level characteristics in a wide-band mobile radio channel. *IEEE Transactions on Vehicular Technology*, 1994, 43(3): 480–486.

[77] Molisch A F, Greenstein L J, Shafi M. Propagation issues for cognitive radio. *Proceedings of the IEEE*, 2009, 97(5): 787–804.

[78] Rubio L, Reig J, Cardona N. Evaluation of Nakagami fading behaviour based on measurements in urban scenarios. *International Journal of Electronics and Communications*, 2007, 61(2): 135.138.

[79] Suzuki H. A statistical model for urban radio propagation. *IEEE Transactions on Communications*, 1977, 25(7): 673–680.

[80] Hashemi H, Mcguire M, Vlasschaert T, et al. Measurements and modeling of temporal variations of the indoor radio propagation channel. *IEEE Transactions on Vehicular Technology*, 1994, 43(3): 733–737.

[81] Schuster U G, Bolcskei H. Ultrawideband channel modeling on the basis of information-theoretic criteria. *IEEE Transactions on Wireless Communications*, 2007, 6(7): 2464–2475.

[82] Wyne S, Singh A P, Tufvesson F, et al. A statistical model for indoor office wireless sensor channels. *IEEE Transactions on Wireless Communications*, 2009, 8(8): 4154–4164.

[83] Mao X H, Lee Y H, NG B C. Statistical modeling of signal variation for propagation along a lift shaft. *IEEE Antennas and Wireless Propagation Letters*, 2010, 9(PP): 752–755.

[84] Abdi A, Tepedelenlioglu C, Kaveh M, et al. On the estimation of the K parameter for the rice fading distribution. *IEEE Communications Letters*, 2001, 5(3): 92–94.

[85] Chen Y, Beaulieu N C. Maximum likelihood estimation of the k factor in Ricean fading channels. *IEEE Communications Letters*, 2005, 9(12): 1040–1042.

[86] He R, Zhong Z, Ai B, Ding J, Yang Y. Propagation measurements and analysis of fading behavior for high speed rail cutting scenarios. *IEEE GLOBECOM, December 3–7, 2012, Anaheim, CA, USA*. Piscataway, NJ: IEEE Press, 2012: 5237–5242.

[87] Chen B, Zhong Z, Ai B, et al. A geometry-based stochastic channel model for high-speed railway cutting scenarios. *IEEE Antennas and Wireless Propagation Letters*, 2014, 14(PP): 851–854.

[88] Ai B, Cheng X, Kürner T, et al. Challenges toward wireless communications for high-speed railway. *IEEE Transactions on Intelligent Transportation Systems*, 2014, 15(5): 2143–2158.

[89] He R, Zhong Z, Ai B, et al. Measurements and analysis of propagation channels in high-speed railway viaducts. *IEEE Transactions on Wireless Communications*, 2013, 12(2): 794–805

[90] He R, Zhong Z, Ai B, et al. Propagation measurements and analysis for high-speed railway cutting scenario. *Electronics Letters*, 2011, 47(21): 1167–1168.

[91] He R, Ai B, Wang G, et al. High-speed railway communications: From GSM-R to LTER. *IEEE Transactions on Vehicular Technology*, 2016, 11(3): 49–58.

[92] Yacoub M. Nakagami-*m* phase-envelope joint distribution: A new model. *IEEE Transactions on Vehicular Technology*, 2010, 59(3): 1552–1557.

[93] Beaulieu N, Saberali S. A generalized diffuse scatter plus line-of-sight fading channel model. *IEEE ICC, June 10–14, 2014, Sydney, NSW, Australia*. Piscataway, NJ: IEEE Press, 2014: 5849–5853.

[94] Salo J, El-Sallabi H, Vainikainen P. Statistical analysis of the multiple scattering radio channel. *IEEE Transactions on Antennas and Propagation*, 2006, 54(11): 3114–3124.

[95] Bandemer B, Oestges C, Czink N, et al. Physically motivated fast-fading model for indoor peer-to-peer channels. *Electronics Letters*, 2009, 45(10): 515.517.

[96] Zhang B, Zhong Z, He R, et al. Measurement-based multiple-scattering model of small-scale fading in high-speed railway cutting scenarios. *IEEE Antennas and Wireless Propagation Letters*, 2017, 16(1): 1427–1430.

[97] Molisch A F. A generic channel model for MIMO wireless propagation channels in macro- and microcells. *IEEE Transactions on Signal Processing*, 2004, 52(1): 61–71.

[98] Salo J, Ei-Sallabi H, Vainikainen P. Impact of double Rayleigh fading on system performance. *2006 1st International Symposium on Wireless Pervasive Computing, January 16–18, 2006, Phuket, Thailand*. Piscataway, NJ: IEEE Press, 2006: 1–5.

[99] Zhou T, Tao C, Liu L, Tan Z. Ricean *K*-factor measurements and analysis for wideband radio channels in high-speed railway U-shape cutting scenarios. *2014 IEEE 79th Vehicular Technology Conference (VTC Spring), May 18–21, 2014, Seoul, South Korea*. Piscataway, NJ: IEEE Press, 2014: 1–5.

[100] He R, Zhong Z, Ai B, et al. Short-term fading behavior in high-speed railway cutting scenario: Measurements, analysis, and statistical models. *IEEE Transactions on Antennas and Propagation*, 2013, 61(4): 2209–2222.

[101] Ephraim Y, Merhav N. Hidden Markov processes. *IEEE Transactions on Information Theory*, 2002, 48(6): 1518–1569.

[102] Gan M, Czink N, Castiglione P, Oestges C, Tufvesson F, Zemen T. Modeling time-variant fast fading statistics of mobile peer-to-peer radio channels. *2011 IEEE 73rd Vehicular Technology Conference (VTC Spring), May 15.18, 2011, Yokohama, Japan*. Piscataway, NJ: IEEE Press, 2011: 1–5.

[103] Vinogradov E, Joseph W, Oestges C. Measurement-based modeling of time-variant fading statistics in indoor peer-to-peer scenarios. *IEEE Transactions on Antennas and Propagation*, 2015, 63(5): 2252–2263.

[104] Erceg V, Soma P, Baum D S, et al. Multiple-input multiple-output fixed wireless radio channel measurements and modeling using dual-polarized antennas at 2.5 GHz. *IEEE Transactions on Wireless Communications*, 2004, 3(6): 2288–2298.